The New Frontiers of Space

Stefania Paladini

The New Frontiers of Space

Economic Implications, Security Issues and Evolving Scenarios

palgrave
macmillan

Stefania Paladini
Birmingham City University
Birmingham, UK

ISBN 978-3-030-19940-1 ISBN 978-3-030-19941-8 (eBook)
https://doi.org/10.1007/978-3-030-19941-8

This Palgrave Macmillan imprint is published by the registered company Springer Nature Switzerland AG
The registered company address is: Gewerbestrasse 11, 6330 Cham, Switzerland

Acknowledgements

This work has been written consulting exclusively materials and data in public domain, available on numerous government websites and organisations. While I have chosen not to include a separate bibliography at the end of the book for reason of space, I have made any effort to cover them in the footnotes, to facilitate autonomous research. As a general rule, I have made frequent use of NASA extensive documentation and consulted all the unclassified sources listed in the EU Commission, JAXA, UKSA and ESA repositories. The sections relative to private groups mentioned in the text (such as SpaceX and Ariane Group, for instance) have made use of freely available company data. The legal section made direct reference to the texts of international treaties and national space laws available on official websites (such as UNOOSA), but they are meant to be of informative use and not for legal advice. All images used have been sourced from stock image sites free for commercial use, otherwise proper attribution has been given. All the tables have been my own elaboration on third-party data, generally from official sources when not otherwise specified in caption. All mistakes are of course mine.

Contents

Acronyms

AEB	Agência Espacial Brasileira
ASAT	Anti-Satellite Weapon
ASI	Agenzia Spaziale Italiana
CASC	China Aerospace Science and Technology Corporation
CASIC	China Aerospace Science and Industry Corporation
CBERS	China-Brazil Earth Resources Satellite Program
CNES	Centre National d'Études Spatiales (France's Space Agency)
CNSA	China National Space Administration
CSA	Canadian Space Agency
CuSat	Cube Satellite
DLR	Deutches Zentrum für Lüft und Raumfahrt (German Aerospace Center)
ECMWF	The European Centre for Medium-Range Weather Forecasts, EU Agencies
ESA	European Space Agency
ESP	European Space Policy
ETS	Earth-to-Space
EUMETSAT	European Organisation for the Exploitation of Meteorological Satellites
FAA	Federal Aviation Administration

FAI	Fédération aéronautique internationale/World Air Sports Federation
GEO	Geostationary Orbit
GNSS	Global Navigation Satellite System
ICBM	Intercontinental Ballistic Missiles
INTA	Instituto Nacional de Tecnica Aerospacial (The Spanish space agency)
ISRO	Indian Space Research Organisation
ISS	International Space Station
ITAR	International Traffic in Arms Regulation
ITU	International Telecommunication Union
JAXA	Japanese Aerospace Exploration Agency
LEO	Low Earth Orbit
MEO	Medium Earth Orbit
NASA	National Aeronautics and Space Administration
NTSI	Non Traditional Security Issues
OECD	Organisation for Economic Cooperation and Development
PPP	Public Private Partnership
QZSS	Quasi-Zenith Satellite System
SASTIND	State Administration for Science, Technology and Industry for National Defense (China)
SEZ	Special Economic Zones
SIA	Satellite Industry Association
STEW	Space-to-Earth Weapon
STS	Space-to-Space
UKSA	United Kingdom Space Agency
UN	United Nation
UNCOPUOS	United Nations Committee on the Peaceful Uses of Outer Space
UNOOSA	United Nations Office for Outer Space Affairs
USRR	Soviet Union
WMO	World Meteorological Office
WTO	World Trade Organisation

List of Figures

List of Tables

List of Case Studies

Chapter 8

Chapter 9

1

Introduction

The overall international space context is changing fast, and a brave new world of challenges and opportunities is opening up. "Competition is increasing; new entrants are bringing challenges and new ambitions in space; space activities are becoming increasingly commercial with greater private sector involvement; and major technological shifts are disrupting traditional industrial and business models in the sector, reducing the cost of accessing and using space. The combination of space data with digital technologies and other sources of data open up many business opportunities".[1] In the context of the above-mentioned quote, the EU Commission was making references to the EU member states; yet, these words can be safely extended to all countries.

As the flight of Wright brothers marks the beginning of the aviation age, the Sputnik is considered the beginning of the space age, to which this book made reference in various occasions. Incidentally, more time has lapsed now from the start of the space age than between the Wright brothers and the Sputnik; yet, due to the formidable challenges of the medium and the nonlinear evolution of the sector that requires a staggering amount of investments to make progress, the space industry itself can still be considered in its infancy.

S. Paladini, *The New Frontiers of Space*,
https://doi.org/10.1007/978-3-030-19941-8_1

If there is something that can be caught at first glance when looking at the space sector as a whole in the last sixty years is that the model for space exploration was based on the direct involvement of governments and the military and funnelled through ad hoc space agencies. It is only in time that a competing approach has slowly emerged, towards a model of public–private partnership entailing a division of labour between the two. The following chapters try to shed light on this complex matter, and they are broadly organised as follows.

Chapter 2 presents a snapshot of the world space sector in the twenty-first century, in terms of its economic dimension and the traditional actors, the nation states, but also discussing a private sector increasingly important in all segments of the space industry. It examines the model that has consistently led the pursuing of space policies, the creation of a national space agency. It is a decade-old model that looks still in favour today, as the recent institution (July 2018) of the Australian Space Agency demonstrates well.

Chapter 3 deals with the space industry analysed under its specific business characteristics, to highlight what makes the space sector an industry in its own right, but also to show that the usual business models can still be applicable, mutatis mutandis. The chapter shows how the business case analysis and project management principles can work in this specific context. It also covers the structure of the sector under its traditional upstream/downstream structure, while leaving the considerations about a more modern, value-chain approach to the last two chapters.

Chapters 4 and 5 outline one of the aspects that have changed the most in the last two decades, the geography of space and its core players. More specifically, Chapter 4 covers Europe, both as a transnational body (the EU and ESA are the obvious topics here) and as the most relevant states in the space business, while Chapter 5 is devoted to the rest of the world, highlighting some of the most important players—USA, Japan, China and India. A brief conclusive section includes the forerunners and the emerging countries that are likely to join the space club in the near future. Each country profile contains a brief sector description, its traditional domains (national space agencies and their links with the domestic military complex) and modern, business-oriented aspects (satellites, imageries, ISS, outer space, launching facilities, ancillary services). It also covers the specific characteristics of each country's space sector, which is at times peculiar, as in the case of China, Japan and India.

Chapters 6 and 7 should be read together since they're both making reference to the same set of international laws and treaties as a regulatory framework. Said that, Chapter 6 treats more specifically security issues, both traditional and non-traditional. The space race started in the 1950s as a military endeavour with national security purposes, carried out by the defence sector. This was the birthmark of the space sector, which is still evident especially in those countries that have joined that race later (even though exceptions do exist). The chapter analyses how wide the security concept can be extended into space and in which way international laws have shaped (or not) space weapons in comparison with nuclear arms and IBMC. The final section is devoted instead to a different kind of security risks, space debris and environmental contamination.

Chapter 7 deals instead with space treaties and laws considered in their commercial and civilian conceptions, showing how this framework has changed in time to adapt to a fast-evolving environment. It is a fact that, while the private sector is now an important actor in space economy, international and national laws disciplining it are severely lacking. If commercial activities such as space tourism, space mining and crowdsourcing in the space sector have to become mainstream, many legal hurdles must be addressed and solved.

Chapter 8 discusses the ongoing trends that will likely shape the space sector in twenty-first century, such as the new generation of telescopes (giant telescopes as much as space telescopes), nano-satellites like the CuSat, and the sector's new segments—space mining and space tourism—that now look on a growing trajectory. Finally, the chapter highlights one perspective often overlooked in books about space, the so-called citizen space. It is no novelty that, after a couple of decades of dwindling interest in space explorations after the hype of the Apollo missions, people are now getting again involved, as the example of ESA Rosetta proved. Initiatives like the World Space Week, established in 1999, are a good example. The UN-sponsored event has grown into the largest public space initiative on the planet, with more than 3700 events in 80 countries in 2017.

Chapter 9 makes a recollection of the space exploration of the Solar System until today and also explores the most visionary and futurist part of space activities, whose borders blur into science fiction. There is the possibility that in fifty years' time humankind will live in a different world, where artificial, asteroid-size satellites with manned stations orbit the planet, private-funded missions are at work to establish the first colony on another Solar System body—Mars or, less likely, the Moon—and compelling evidence of alien (albeit not intelligence) life has been found in the Earth's neighbourhood (e.g. Europa's oceans or Titan's methane seas).

A brief section of conclusions (Chapter 10) deals with already planned for the year just started (2019) and more in general for the next decade. The 2020s look astonishing in terms of number and quality of the missions ahead, one of them among all: the search for alien life. But not all scenarios look equally promising. Among all those positive perspectives highlighted above, there are threats looming ahead and, apart from the most obvious ones, there are others that are not exclusively linked to space exploration—such as climate change and security issues. The already mentioned US Space Force is only one in a general trend that can reproduce in space a dangerous geopolitical instability that exists on planet. The growing problem of debris is another, and it is an excellent example of how, in all the so-called commons—oceans, atmosphere and forests—cooperation and a strong, internationally recognised legal framework is paramount to ensure a bright future.

The intended audience of this book is both practitioners in the field and scholars of international relations, economics, business and security studies. There are many books on space available, especially at this crucial moment, and this one intends to contribute to two aspects.

One is the thematic extent. Much of the present literature does not explore brand-new concepts (say, asteroid mining, or the plans for lunar bases) for the simple reason that they were topics virtually non-existent just a few years ago, and they're still the domain of specialised press or academic articles.

The other is the interdisciplinary character of the project. The majority of the books already in the press are written from the point of view of traditional security or from a business-economic perspective, but not considering both. However, a trans-disciplinary approach is essential in the case of the space sector, especially if considered in a future perspective, given the inescapable security aspects and the obvious economic implication of any initiative.

There are two more features to highlight. The first one is that, while there is an obvious progression across the chapters, first examining what there is on planet Earth under different points of analysis and then moving to the Earth's orbit, the Solar System and farther on, the chapters themselves are meant to be read independently. A few inter-text references and footnotes facilitate this kind of approach, listing obvious connections.

Case studies are the second feature. The rationale is to provide a spotlight to some selected topics, which have become, or are going to become, critical, or, in other cases, that are important to understanding the zeitgeist of the space sector. One, for instance, analyses the way space missions have become everybody's interest and even entered primary schools (ESA Rosetta). Another covers non-traditional security issues, such as the growing dangers of debris and the legal implication of collisions (e.g. the case of the Russian Cosmos 2231 accidentally destroying the satellite Iridium 33). The closing case study explores instead the rich links between science and science fiction, showing how the intersection between the two is wide (scientist-writers are a common feature) and that science fiction concepts often foreshadow real science missions. Moreover, since space does not belong to Western only, the case study engages in at least one alternative vision of humanity in space in Japan's popular culture (the Gundam phenomenon).

For practical reasons, the references for the case studies have been listed separately for the rest of the book. While references that have been used both in the book chapter and in the single case studies are present in both lists, the ones exclusive to case studies have been given in the case study only.

There are a few limitations to acknowledge about this study. A conscious decision has been made of limiting to the bare minimum the technical aspects of a notoriously tech-heavy sector. This has been a difficult choice, because it has also meant to overlook a series of relevant topics—such as the technical details of SpaceX's innovative approach and factors related to space exploration. The author is well aware of this issue. However, if there is a trove of freely available information on the web for tech-savvy space lovers, there are not enough books that cover what can be defined "the soft side of space", which addresses its social, economic and political aspects. This is the gap the present book is trying to fill, offering only a sketch of the more technical aspects together with suggestions for further reading.

The second limitation is its geographical cover. A quick look at the table of content shows that not all the countries got an equal share in terms of depth of analysis. While the predominance of the USA and the novelty of an enhanced China's presence are responsible for the extended coverage given to them, the other spotlight available has been granted to Europe (intended as ESA and selected few EU countries). The rest of the world, even including important space nations such as Japan, Brazil, Canada and India, have been given only limited coverage. Again, this has been a conscious decision, due to sheer reasons of space. There is also a specific reason for having an entire chapter devoted to Europe, and it is the fact that the EU space sector is now going through turbulent times. This has been only in part due to the Brexit case, even though Brexit itself is a cautionary example of the importance of a regulatory framework.

The third and final limitation is that what represents the most interesting, visionary part of the space exploration remains at the margin of this monograph, and it has only been summarily outlined in Chapter 8. Any in-depth analysis of this subject will take at least a book on its own, if not more. This book can only suggest what are the possible outcomes in the medium term, discuss the challenges humankind face as a species and provide references to upcoming missions.

Never more than on this occasion, it is right to say the sky is not the limit.

Note

1. European Commission. (2016). *Final Communication from the Commission to the European Parliament, the Council, the European Economic and Social Committee and the Committee of the Regions. Space Strategy for Europe.* COM 2016/705.

2

A Snapshot of Space: 2018

2.1 The State and Size of the World's Space Sector: A Snapshot

An estimate of the size of the world's space sector can prove surprisingly difficult to achieve. First of all, it depends on which activities are to be included, and which vary a great deal from a report to another[1] and from country to country. The literature lists four uses of space (Hays et al. 2000), namely:

1. Intelligence and information use (security-oriented activities, not necessarily military but somehow of strategic interest. Surveillance satellites are a good example);
2. Military uses (more tactical than the previous one, generally in the case of conflicts, such as drone strikes);
3. Civilian uses (government-led activities, meteorology, environment, Earth observation);
4. Commercial uses (private-led activities; the majority of the sector's downstream operations are included in this category[2]).

S. Paladini, *The New Frontiers of Space*,
https://doi.org/10.1007/978-3-030-19941-8_2

This is important because there are states—and the USA is one of them—where there is a clear distinction between the space sector for civilian use and what remains military precinct; this distinction has a direct implication on the regulatory framework, as shown later in this book. This is not necessarily the case in other countries, where the influence of the military sector is evident in space design and planning.[3]

The above-mentioned categorisation does not take into account space cooperation between civilian actors (both state and private), which is increasingly common, or cross-country cooperation. However, cooperative efforts must be acknowledged to avoid a duplication of the estimates. The most traditional type is constituted by cooperation over space programmes and science: this requires that the two partners have at least comparable technological advancement to make it work (often both are spacefaring countries); the second is about space navigation, and in this second case, the partners do not necessarily need the same level of technology. This is why countries with even modest technical capabilities can be involved.

Good examples are NASA partnerships with the rest of the world over a series of missions; another is the ongoing cooperation between Europe (both as ESA and single member states) and Japan, which are the results of a long-term history of joint research (Correll and Peter 2005).

With all these caveats, a reasonable estimate will give the 2017 worldwide space market at 383.5 US$ billion (The Space Foundation 2018)[4]; this represents a substantial increase over 2016, which was estimated at 329.3 US$ billion. A more conservative one is the one reported by Satellite Industry Association (SIA) in its yearly report, which put it as "only" 348 US$ billion.

These figures tell the story of a sector growth that is accelerating: in the Space Foundation's figures, the increase 2016–2017 was US$6.3 billion. The same two reports were putting 2013[5] figures at $256 billion and $231 billion, respectively.

The forecasts are, however, of an even stronger growth over the following years, up to 587 US$ billion (OECD methodology 2007) or 720 US$ billion (the Space Foundation's estimates) by 2030.[6] In terms of rate, the OECD suggested an average of 5% year-on-year.

Later in this book (Chapters 4 and 5), there is an account of how each country contributed to this figure, together with an analysis of somc selected countries' space sector. Here, it is enough to say that, in terms of state agencies, at least seven of them have spent more than 1 US$ billion on space in 2017, with a private sector that is increasing its share each year. OECD countries still account for the largest space budget at world level, even though an increasing amount of space activities now takes place outside the OECD (Table 2.1).

As for 2018, the USA still detains the lion share in the world space sector, which the US government spending on space that is about 57% of the world's total.

The American launch operations have been reduced since the NASA retired its last Space Shuttle in 2012. However, the USA remains a powerhouse in space spending nonetheless. For example, in 2013, the USA spent about US$40 billion on space activities in 2013, still more than all other countries combined.

The activities carried out by NASA accounted for US$18 billion, with the bulk of this amount spent on space exploration, research and cross-agency report. Other areas funded by the space budget included activities that fall under several departments including Defence, Energy, Transportation and Interior's Geological Survey.

One of the quickest ways to assess the status of the world's space sector is to have a look at the launch statistics, in terms of frequency, success and, more importantly, operators, private and public. Albeit the dynamism of the sector, there are still only a reduced number of national space agencies that maintain in-house launching capabilities, while others, even having developed them, have either renounced (Canada) or turned to commercial launch providers. An example is the government-funded but still private corporation Arianespace (see Chapter 4), which provides services to a series of agencies, NASA, ISRO and ESA among them (Spaceflight 101 2017). Since 2014, SpaceX has been serving as one of NASA's carriers for resupplying missions to the ISS (NASA 2016), and other private entities have entered the arena, as illustrated in Chapter 5 when talking about China.

An interesting comparison could be done by looking at 20 years ago. A snapshot of 1998[7] will look quite different, even though it is the case

Table 2.1 World's government expenditure for space programmes (2016) (in million US$)

United States	35,957
Canada	434
EU	1929
Ireland	30
UK	743
France	2792
Spain	293
Portugal	23
Italy	945
Germany	1984
Luxembourg	78
Denmark	42
Norway	107
Poland	52
ESA & Eumetsat	4793
Greece	28
Europe	10,372
Switzerland	187
Netherland	159
Belgium	231
Sweden	122
Austria	81
Russia	3182
Kazakhstan	45
Ukraine	25
Belarus	83
China	4909
Japan	3018
South Korea	671
Taiwan	75
Bangladesh	21
India	1092
Indonesia	160
Pakistan	32
Australia	235
UAE	376
South Africa	20
Saudi Arabia	127
Israel	155

(continued)

Table 2.1 (continued)

Iran	109
Venezuela	163
Argentina	129
Brazil	163
Mexico	145
Bolivia	38
Total	62,200

Source Author's elaboration on Euroconsult, 2017 and official agencies data, 2018

to mention that a rocket payload is as relevant as the number of launches (e.g. the same spacecraft nowadays may lift more than one satellite).

The satellite industry is by far and large the dominant segment, accounting about 77% in 2017 (SIA 2018). As a whole, the number of countries operating a satellite has kept growing over the years (59 countries operate at least one of them, alone or in a consortium, while the US alone operates more than 600). The satellites constitute the majority of the rocket payloads. They are generally classified as military or civilian, even though in recent times there have been important innovations in their very conception, such as the CuSat (see Chapter 8).

Historically speaking, satellites from the military sector have been the first ones to be put into orbit, followed in the middle of 1970s[8] by civilian satellite programmes from a growing number of states, changing the geopolitics of space also under a more security-related point of view (which will be discussed in Chapter 6 in detail).

Due to its importance, the satellite industry has been treated in detail in Chapter 3, providing figures and dealing with the classification of satellite applications by end-use.

Here, it is enough to say that, while telecommunication is the one enjoying the most impressive growth in terms of revenues in the last decade, it is the navigation system that plays a critical role in the satellite business.

In this specific area, the proliferation of providers witnessed in the other areas has not manifested yet. In the navigation system—otherwise called Global Navigation Satellite System (GNSS)—the US-provided GPS was until recently in a regime of virtual monopoly.

As of 2018, there is only a handful that dominates the sector:

- the U.S. Global Positing System (GPS);
- the Russian Globalnaya Navigazionnaya Sputnikovaya Sistema (GLONASS);
- China's BeiDou Navigation Satellite—System (BDS, afterwards called Beidou);
- the EU's Galileo Satellite Navigation System (Galileo).

There are also a few regional systems, which, differently for the ones above mentioned, only serve a restricted area. This is, for example, the case of Japan's own regional system, the Quasi-Zenith Satellite System (QZSS).

The four global systems are not, however, at the same stage of development.

The GPS and GLONASS have been the longest in operation and provide worldwide coverage; Galileo is still catching up after a series of delay; and BDS has been providing positioning, navigation and timing (PNT) service to the Asia-Pacific region since December 2012 and it is supposed to offer global services by 2020.

The implication of Beidou has been examined in the case study in Chapter 5, as the system represents a formidable tool of diplomacy China deploys in its BRI initiative around the world.

How the world got to this stage in sixty years is discussed in the following section.

2.2 A Brief History of the Space Race

The world's first artificial satellite, Sputnik 1, was launched on 4 October 1957, changing the world forever and beginning what it is now defined "the space era". About five years earlier, in March 1952, the concept of weapons in space had been introduced to the public through an issue of Collier's magazine where Wernher von Braun described an orbiting space station that could possibly host nuclear weapons (Bulkeley and Spinardi 1986). This was not an accident: since the beginning, space exploration has been closely related, if not equivalent, to a weaponisation of the space itself, and therefore an area where states in general and their military sectors in particular were the protagonists.

This also explains a lot of the successive history of space, until the moment commercial satellites first and private companies later on appeared on stage. What kind of implications this had on the current character of the sector, especially in terms of international relations and regulatory framework, has been discussed in Chapters 6 and 7.

The tables in Chapter 9 and Appendix 1 provide a more analytical view of space missions, launches and other statistics, while the following notes intend instead to offer a few highlights and focus more specifically on what happened on planet Earth.

Things have gone a long way when thinking that all started with a small probe of a size of a ball (58 cm in diameter) and the weight of 83.6 kg (Fig. 2.1). The Sputnik orbited the planet for less than two hours (98 minutes to be precise) but it did not matter. It was enough to start that race that would have taken a human, less than twelve years later, to walk on another celestial body's soil. Regarded in this perspective, the achievement was nothing but astonishing.

The USA claimed an earlier start, at least in terms of manned flight.

In September 1956, Iven C. Kincheloe Jr. (US) flew to a height of 38.5 km (126.200 feet) with a Bell X-2 rocket-powered plane, becoming the first human ever to fly above 100,000 feet. While this is far away from what FAI considers the limit of space (see Chapter 8 for a discussion about this point), "Kincheloe was immediately hailed as 'The first of the spacemen' and 'America's No. 1 Spaceman' back in the days and still represents a great achievement in astronautics" (FAI 2016, online).

However, after the Sputnik's launch, the USA decided to commit federal resources into the space race, and in 1958, Eisenhower started the man-in-space Project Mercury; in the same year, NASA was established to look after the nation's space endeavours.

If the 1950s were the decade of probes in space, the 1960s were about men—first launched into orbit (1961, the Russian Gagarin and the US Shepard Jr.) and then to the Moon (1969, Armstrong and Aldrin), while the 1970s saw the creation of orbiting stations (Salyut 1, 1971; Skylab, 1973). The space race ended with the Apollo-Soyuz Test Project in 1975,

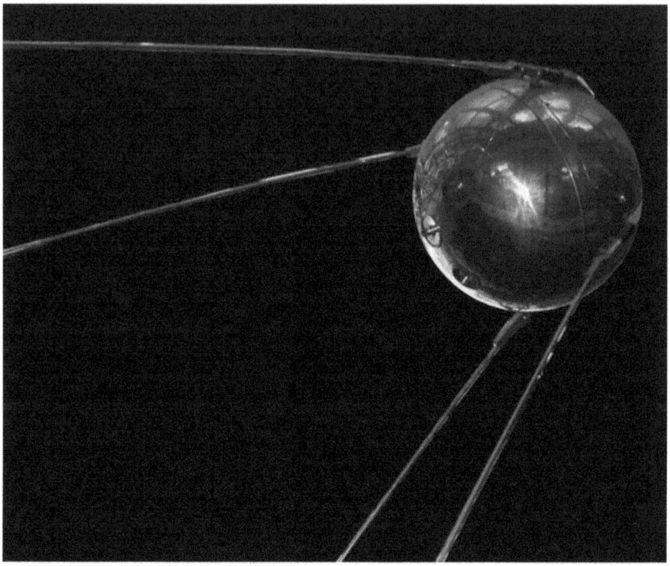

Fig. 2.1 Sputnik 1

paving the way to a less confrontational, non-territorial vision of space that took hold in the following decades.

Other countries followed on the heels of USA and USSR. While it stands uncontested that Japan was the fourth country to get to space, at least four countries claim the third place. Canada built and operated the third satellite to be launched (Alouette 1, 1962), the UK operated the third-ever satellite (Ariel 1) five months before the Canadians but did not build it. Moreover, NASA launched both of them. If launching is what marks the transition to the space age, then this title belongs to France, whose space agency CNES sent the satellite Asterix into orbit in 1965 using a Diamond rocket. Finally, if a manned flight is what that matters, another country, Czechoslovakia[9]can claim the honour with Vladimir Remek, shipped to the orbital space station Salyut 6 in 1978.

The 1980s proved fundamental for the sector, with a series of important events that took place with a revolutionary concept: a reusable spacecraft, managed by a space agency, which could achieve a substantial reduction in the access cost to space. The idea was developed to become the US Space Shuttle.

Few remember today that, as much as celebrated the US Space Shuttle can be, it was not the only one in development in those years. The Soviet's Buran was very much in the race for the perfect shuttle. And while the world only learnt about it on 15 November 1988, after a secret development which went on of a decade and cost billions of dollars, the Buran eventually launched only once, disbanded like many other projects by the collapse of Soviet Union.

If the decade itself was crucial, 1986 is a year that will remain in history for more than one reason.

First of all, the Soviet launched the first module of the space station MIR, which remained in operation until 2005. The MIR was a revolutionary concept, building on the existing experience of the Salyut station and presenting a structure composed of separate modules assembled directly in orbit, like the ISS currently in operation.

But 1986 also saw a tragedy in space, one that affected the Space Shuttle. On 28 January, the Space Shuttle Challenger at its tenth flight broke apart 73 seconds into its flight, killing all seven members and grounding the entire Shuttle fleet for almost three years.

The US Space Shuttle cost more than budgeted and eventually the fleet was retired, also due to safety concerns.

ISS had to be serviced, from that moment on, by the Soyuz spacecraft, while the return to the use of expendable rockets and the increased demand for satellite launches persuaded the US government to provide incentives to the private sector for developing independent launching facilities.

The era of NASA subcontractors had begun.

The 1990s saw, together with the end of the Soviet Union, the first international cooperation between the two former space rivals, with the astronaut Krikalev becoming the first Russian on board on a US spacecraft (during a Shuttle flight) in 1991 and the Russian hosting a US member (Thagard) on the MIR for 115 days. This and the following exchanges paved the way to the ISS in 1998, an international initiative[10] in which both the USA and Russia were founding partners.

The first decade of the new century witnessed an unprecedented expansion in the number of satellites and the widespread use of navigation systems. In what is another watershed of the space era, the word saw in 2011 the last Shuttle mission (Atlantis, landing on 21 July 2011[11]), and the launch, the very next year,[12] of SpaceX's Dragon spacecraft to resupply the ISS.

The space sector is now witnessing another moment of frenzy.

Only the activities beyond Earth's orbit remain the precinct of a selected few (USA, Russia, a handful of European countries led by ESA, Japan, China and India). More and more countries launch satellites into Earth's lower orbit (LEO) and geostationary orbit (GEO), alone or as a "piggy-back" on others' states missions. This is a recognised stage of any country's development programme to become a space-active nation (Peters 2016)[13] and, as mentioned in the previous section, more than 60 countries have taken part to space mission even without having launch capabilities or even fully manufactured the satellite by themselves.

This number is due to grow, creating a new situation, with new actors, new demands and new, formidable challenges. But if there is one thing that has remained constant in sixty years of space activities is the attractiveness of the civilian space agency model, with some early examples created even before the Sputnik. It is worth asking why, and why they are still relevant today in such a different landscape.

The next two sections deal with this particular aspect of the space sector: its actors.

2.3 The Global Space Industry in Evolution: A Composite Sector

A quick look at the space sector will show that *Outer space activities have evolved* significantly over time, and what was previously the exclusive domain of a restricted number of states, now have expanded both in scope and plurality of actors involved. As Tronchetti states, "thanks to technological advances and the easing of governmental restrictions, space activities are carried out on a much larger scale and involve subjects of both a governmental and non-governmental nature" (Tronchetti 2013: 3), a fact that makes space business increasingly profitable and attractive to potential investor. And if until recently these activities have been carried out only by state-owned enterprises (SOEs) or national space agencies, now private companies are more and more represented in all areas, including the ones, like launching, formerly reserved to the government.

This virtual monopoly of the government on space activities, as it will be better explained in Chapter 7, has profoundly affected the way the regulatory framework of space has emerged over the decades; on the other hand, the end of it is already producing dramatic changes. This is nowhere as evident as in the main international treaties dealing with space at some titles, which, Genta explains, "were heavily influenced by the belief that states were the only actors in space and that exploration could be peaceful only if states refrained from claiming 'things' that exist beyond the Earth as their own, and from taking any sort of weapons in space. Everything of value existing there was to be considered as belonging to humankind in general, and should be exploited, if at all, in the interest of all humans" (Genta 2014: 2).

With the multiplication of state players and private companies, the 1967 Treaty of Outer Space, with its provisions of making the national activities in outer space a responsibility of the states (and not of the single companies), is becoming increasingly outdated. Moreover, the treaty is only applicable to the states that have ratified it; some did not, nor have any intention to do so.

In the past, even in those instances in which space activities were outsourced to the private sector, governmental agencies and the military designed and supervised them. Now it is increasingly not the case. There are only two segments where the government is still playing a prominent role: in military-defence applications and space exploration—i.e. beyond Earth's orbit.

The process that led to a more composite space sector began in the 1980s, when a new concept became popular: space agencies were not to deal with all kinds of activities, leaving industrial applications, such as telecommunication satellites, but also meteorological and Earth resources satellites, to private companies, while concentrating on their main business, namely science and space exploration.

Not all private operators in the space industry appeared at the same time, though.

The segment of commercial satellites was the first to make an appearance as early as 1962 (Telstar 1, launched into space by AT&T and Bell Telephone Laboratories), with mobile telecommunication coming afterwards. Private spaceflights, however, is a more recent business, even though state-supported but commercially managed companies (such as Arianespace, the world's first commercial launch service provider) have been in operations since the 1980s. But on the whole it was from the 1990s onwards that totally private companies entered the segment, first of all as NASA subcontractors.

As a whole, the private space sector has grown rapidly over the last two decades, and start-ups have made their appearance, too. According to some estimates, more than 80 companies were established by venture capitalists in between 2000 and 2016, "eight of which have been sold at a total value of US$2.2 billion. Among them, these firms have raised more than US$13.3 billion in investment, with two-thirds of that being raised within the last five years" (Christensen et al. 2016: 23).

In terms of space venture start-ups, 2016 and 2017 were so far the years that recorded the highest investments. 2016 saw more than 100 investors putting about US$2.8 billion into 43 start-ups across 49 deals (FAA 2018). Data for 2017 are not available yet, but are expected to surpass 2016 both in investments and number of deals.

A good way to assess the importance of each segment in the global space industry is to break it down to its main segments and components.[14] In an industry estimated at US$348 billion (SIA 2018), about US$268.6 billion belongs to the satellite segment and its specific subsegments (Fig. 2.2).

They include, as it will be shown in the following chapter, the following segments:

1. Satellite manufacturing;
2. Satellite applications; this represents by far the biggest subsegment of the sector, which has witnessed impressive growth over the years. The segment can be further split into three subsegments:

    2.1 *Satellite communications.* Historically, it is the most important, since Telstar was launched in 1962. Intelsat I followed in 1965 and took its place in GEO, which became the most coveted orbit to locate

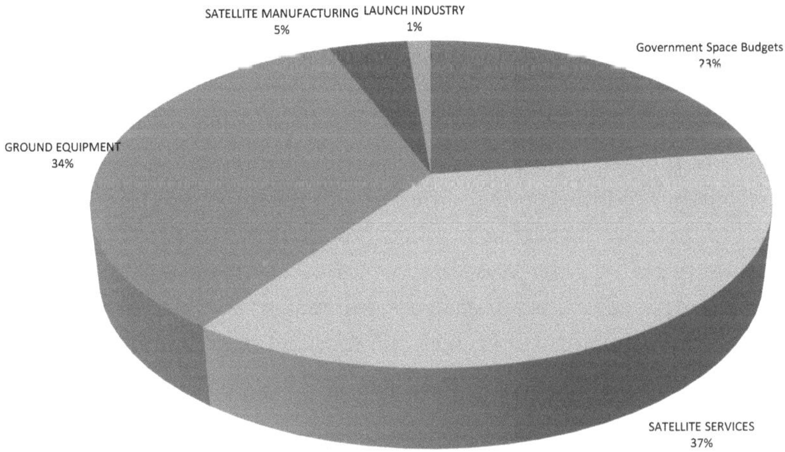

Fig. 2.2 The world satellite industry in 2017 (*Source* Author's elaborations on SIA (2017)) (US$ billion 347.9)

communication satellites. But it was since the 1990s that the sector knew a real explosion, with the dot.com revolution that put up hundreds of satellites in LEO and saw mobile and broadband companies—such as Iridium, Globalstar, ICO, and Teledesic, just to mention a few—making their appearance.

2.2 *Earth observation (EO) applications, also called remote sensing.* This is an area that, while still earning less compared to others, is one of the most promising, especially in the context of climate changes and increasing weather instability. This is of special interest to emerging economies for the help they can provide to agriculture and disaster management.

2.3 *Global navigation systems.* As mentioned in Sect. 2.1, this specific subsegment is of strategic importance on top of its commercial and civil applications, even though a substantial part of the revenues is given by the receivers. The devices themselves, it must be noted, are included under another entry, statistic-wise.

3. Ground equipment (satellite navigation equipment, GNSS receivers; traffic information systems; aircraft avionics, maritime, surveying and rail); this is what is called downstream.

A segment somehow separates even though connected to the satellite business is space operations, i.e. the launching industry. The remaining US$79.3 billion is divided between government space budgets and human commercial spaceflight.

There is another segment, not taken into account in SIA reports, and which partly account for the different estimates between SIA and the Space Foundation. It is what is called "ancillary Services" or "specialised support services". Under this category, the UK Space Agency includes "launch and satellite insurance (incl. brokerage) services, financial and legal services, software and IT services, market research and consultancy services, business incubation and development, policymaking, regulation and oversight" (UKSA 2018).

The figures of these segments have been discussed in more details in Chapter 3. As a whole, telecommunications system sales seem to have peaked between 2005 and 2010 and grown at a slower pace after 2011. In some areas of the world—such as Europe—they have decreased since 2013. However, there is a growing demand for broadband and satellite communication markets especially in areas without cables, such as emerging economies. While their numbers are still limited at the moment, they are expected to grow significantly in the next decade.

Finally, there are other, not fully accounted segments, which promise to become increasingly important in the space industry: space mining and space tourism. Both segments have been examined into further details in Chapter 8 in terms of their concept, history and future development. Here, it is only the case to give some industry data.

Space tourism is a restricted market so far, both in terms of actual demand (compared to the potential, which is much higher. Nowadays, only very few can afford the costs) and limited offer. As of 2018, space tourism is still in its infancy; only orbital space tourism has taken place (so far) provided by Space Adventure/Soyuz to a handful of ISS visitors, whose travel ticket is quoted in terms of US$20–35 million. Sub-orbital flights[15] continue to be tested, with Blue Origin and Virgin Galactic on the lead and already collecting reservations[16] (the ticket is about US$250,000). Once the countries' regulatory framework, and not only the technology, will be apt to market requirements, space tourism is forecasted to grow fast, with the USA as the leading market.

In terms of expected revenues over the next five years, estimates (*Reuter*, 17 August 2018) give the sector a 17.3% CAGR, going up from US$490 million in 2017 to US$1270 million by 2023. Moreover, a 2010 report from FAI estimated a space tourism growth up to a billion-dollar market within 20 years.

On the other hand, space mining has already seen a few companies around the world willing to partner with governments and launch their bids to NEOs' raw material retrieval. The challenges here are most technical as the legal framework is evolving in a fast way to allow the exploitation itself.

The forecast gives here a growth from US$0.65 billion in 2018 to US$2.84 billion by 2023, with a CAGR of 23.6%. Albeit the Europeans lead in this segment, due to the presence of a favourable legal framework and investment conditions, 75% of the active companies[17] are US-incorporated. Some of them as Deep Space Industries and Planetary Resources are, however, based in Luxembourg.

It is worth noting that, different from space tourism, space mining sees the presence of many space agencies, actors whose characteristics and mission have changed a great deal over the years, as the following section shows.

2.4 National Space Agencies: Old Concept, New Aims

NASA needs no introduction,[18] even among the non-specialists.

And yet, the entire concept of space agency is somehow less evident. Not all the countries set up a devoted (generally civilian) agency to look after the space programme, for whatever reason it might be. Often the military interests were predominant, and prevented this to happen; in other cases, the various ministries in charge of portions of the space programme refuse to give up their competencies and retain influence.

In this second case (Verger et al. 2003), space competencies are handled directly by a specific ministry (education, research and technology; industry or trade; defence) or by a composite, inter-ministerial entity, often a committee. A recent report, *A Roadmap for Emerging Space States*, by the

International Space University, looks at it as at a three-way solution: a state can organise its space endeavours as a centralised effort, as a decentralised one or as some combination of the two (ISU 2017).

A space agency can either act as an operator, taking an active part to some or all of the activities, such as science, research or a coordinator, putting together and organising the efforts by a plethora of actors, private and public. And if Spain's INTA,[19] the oldest space agency still in business today, is a typical example of first kind, Luxembourg Space Cluster is clearly only coordinating other's efforts. Interestingly, an agency can start as an operator and adding over the time coordination functions in selected areas of its activities. This is the case of NASA, which, when the Space Shuttle went out of service, contracted out launches to commercial entities such as SpaceX (Pearlman 2011).

In terms of the history of the agency model, their creation has progressively emerged as the dominant trend in the space sector and NASA, established as early as in 1958, has influenced all the followers in terms of structure, prerogatives and willingness to engage potential partners in cooperative ventures.

Attractive as the NASA model could have been; however, it was necessary to reach the end of the Cold War, and a new geopolitical equilibrium before the model itself could spread out beyond the USA and its NATO allies. This is one of the reasons why, from the 1990s onwards, the world saw the multiplication of civilian space agencies.[20]

Why does a country decide to set up a space agency, anyway? As briefly mentioned above, it is fair to imagine that not everybody in a country's existing administration might look favourably to an independent body. Still, the reasons can be many; some countries—South Africa is one of them—see the space agency as a way to support the country's overall economic growth (Andsell et al. 2011; Mosteshar 2013).

The idea is not far-fetched; there is abundant literature about the multiple benefits coming from the space sector, economic (Wood and Weigel 2014) as well as technological (Paikowsky 2017). This is the case even when the rate of returns (calculated by some scholars to be around three times the investments; Cohendet 1997) is not always immediate and "generally happen over time across different parts of the economy. Since the 1970s, industry has used space applications to create new business solutions. Agricultural, maritime, and consumer technologies are just a few of the areas using space applications" (ISU 2017: 18).

Others—and the USA is among them—do it for space exploration first and foremost (Dunbar 2017; Extance 2014). But there might be other, equally valid reasons for a state to engage in space activities, and Logsdon (2005) mentions a few ones, such as: providing benefits to society; increasing national prestige (UAE); gaining scientific knowledge; enhancing national security (Malaysia and Vietnam), at times due to non-traditional threats like typhoons (Philippines); and inspiring younger generations.

Nothing shows the different motivations for a country to engage in a space programme better than the attitude towards the development of domestic launch capabilities. The prestige of belonging to a "space club" is certainly a factor for some; others, even when they technically can, choose not to pursue it. A well-known example is Canada: the country became in 1962 the third to launch a satellite into space (Alouette 1) and the first to have its own satellite for geostationary communication. But today, Canada has not independent launching capabilities[21] and uses other vectors (e.g. SpaceX) to send its satellite into orbit.

Not all space agencies have been equally successful. Some of them have gone through extensive restructuring and change of name; others have raked up failures.

As discussed in the ISU report, there are a few things that help in the road to a sustainable space programme; a strong connection with the domestic (and, as it will be shown on the next section, international) space sector is one, as it is a well-established presence on the national territory, with multiple locations for research centres and launching sites. Again, NASA represents a textbook example.

Most importantly, as Weeden clarifies, "decisions that are made by individual government agencies or entities without coordination and input from other stakeholders, including the private sector, are likely to be suboptimal. This is because barriers between commercial, civil, and national security space activities are increasingly becoming blurred" (Weeden 2017).

Moreover, a series of government measures can be taken to encourage and nurture the domestic space industry, with tax relief, incubators and specific regulations for industrial clusters, as it has been done recently in

Luxembourg. Budget allocation is another important component in the success of space agency's activities and, in general, of the space sector of a state.

There are in this area a few options open to the government to finance these activities. Direct funding is the most obvious and traditionally adopted way, but others do exist to support efforts. How often funding is revised and incremented if the case is a crucial factor, considering how peculiar and resource consuming the space industry is (a detailed analysis appears in Chapter 3), and there are pros and cons[22] in annual (NASA) or multi-annual (ESA) deliberations about space budgets.

The following tables provide comparative data from selected space agencies[23] (Table 2.2).

It is important to notice the countries' fund allocation to the space sector against the overall GDP.[24] Space budgets are generally small (only a few of them, the USA and Luxembourg for example, have a share larger than 0.50% of the GDP), often dwarfed by allocations to defence or other sectors of the economy. Space agencies taking an active role, such as NASA, require more funding than the ones that only coordinate others' efforts.

Transparency about the allocation of funds is another essential factor to take into account, especially in the case of joint initiatives with the private sector (Brennan and Vecchi 2011), and NASA is a reference to every agency in terms of a transparent and well-accounted for management.

A final trend to be mentioned is the dynamic of increasing cooperation that is emerging both among the space agencies themselves and between them and the variety of new partners now available both at national and international level. Peter, who devoted a study on this important aspect, observed that both the multiplication of space actors and the removal of "inter-bloc" cooperation barriers are dramatically changing the geoeconomics and geopolitics of the space missions in a fashion impossible only 15 years ago. "New axes of cooperation are arising, some are deepening, while others are weakening, indicating that the upcoming years will undoubtedly see new alliances that will reflect the emerging 'post-post-cold war' geopolitical context. Moreover, as many successful examples illustrating the benefits of working together with others to achieve common goals exist, states newly involved in space activities find cooperation helpful in mitigating the inevitable postpartum stress and risks associated

Table 2.2 National space agencies

Country	Name	Creation date	Selected budget data 2017[a]	Launch capabilities
Americas				
Argentina	CONAE	1991		
Brazil	AEB	1994	260	
Canada	CSA	1989	488	X
Mexico	AEM	2010		
Uruguay	CIDA-E	1975		
USA	NASA	1958	19300	X
Europe				
Austria	ALR	1972		
Belgium	BIRA	1964		
Europe	**ESA**	1975	6425	X
France	CNES	1961		X
Germany	DLR	1969		
Greece	ISARS	1555		
Hungary	HSO	1992		
Italy	ASI	1988		X
Luxemburg			223	
Norway	NSR	1987		
Russia	ROSCOSMOS	1992		X
Spain	INTA	1942	170	
The UK	UKSA	2010	414	
Turkey	TÜBITAK	1985		
Ukraine	NSAU	1992		X
Africa				
Nigeria	NASRDA	1998		
South-Africa	SANSA	2010	19	

(continued)

Table 2.2 (continued)

Country	Name	Creation date	Selected budget data 2017[a]	Launch capabilities
Asia-Pacific				
Australia	ASA	2018		
Bangladesh	SPARRSO	1980		
China	CNSA	1993	4300	X
Kazakhstan	NSA	1991		
India	ISRO	1969	1100	X
Iran	ISA	2003		X
Israel	ISA	1983		X
Japan	JAXA	2003		X
Malaysia			8	
Singapore	CRISP	1995		
South-Korea	KARI	1989		X
UAE	UAESA-MBRSC	2014	150	
Vietnam	VAST-STI	2006		

[a]In million of US$

Source Author's elaboration on the countries' national budget 2018–2019; OECD (2017)

with space initiatives. Furthermore, all recent newcomers in space seek to establish cooperative agreements with spacefaring countries to benefit from potential transfers of technology and tacit knowledge" (Peter 2016: 5)

This specific aspect—cooperation efforts—is going to be explored into further detail in the next section.

As a final note, it is fair to say that the agency model keeps being attractive.

The proof is that the last space agency has been created as recently as July 2018 (the Australian Space Agency[25]) and more can be expected in this sense in the following years.

2.5 The Twenty-First Century: A Cooperative Space

Another trend worth noticing, and which stems directly from the new dynamism in terms of private space actors and national agencies of the last two decades, is the growing number of bilateral and multilateral agreements, both at the level of states and of the single space agencies. Chapters 5, 6, and 7 mention cases where cooperation has produced substantial results. The most obvious is, of course, the ISS, to which the case study of Chapter 6 is devoted, but the numerous space missions ongoing at the moment are another example.

And while it is true that the space sector has oscillated between conflict and cooperation during its long history, and that the competition between the two superpowers has been the driving force behind space exploration (Machay and Hajko 2015), it is also true that nowadays, compared to the days of the Apollo programme, the focus has shifted instead towards a more cooperative approach.

There is a reason for that: the benefits of international agreements, such as specialisation and cost sharing. Cooperation is therefore driven by a twofold motive: economic and strategic.

There are other positive returns. At a domestic level, cooperative initiatives increase space missions' political sustainability, especially when those missions are expensive and relying on yearly negotiated budgets, as highlighted in the previous section.

This might prove fundamental for the life of the missions themselves. "Once cooperation has commenced, cancelling a program becomes inconsistent with political sustainability as long as the utility cost associated with the loss of diplomatic benefits and the negative effects on reputation of terminating an international agreement is larger in magnitude than the utility cost that must be paid to maintain the system" (Broniatowski et al. 2006, online).

Reducing costs by sharing initiatives has been on space agencies' agenda for a long time. This course of action has been, for instance, explicitly suggested to NASA, more than twenty years ago, by the US Congressional Budget Office (CBO) in a report that highlighted two areas of improvement in the agency's management: the necessity of reducing high fixed costs and dealing with funding coming late in a project's life cycle (CBO 1994).

Those were still early days, where opportunities for cooperation were more limited in terms of actors and initiatives. Now, with more than 50 countries launching satellites into space (The Space Review 2018, online), and many private companies engaging in such an array of diversified efforts (see Chapter 8), this comes out as a more feasible option.

An example of the success of global efforts in space cooperation (Haynes 1987) is the Deep Space Network (DSN), the largest scientific telecommunications system in the world, under the joint direction of IND (Interplanetary Network Directorate), NASA-JPL (Jet Propulsion Laboratory) and sets of antennas in Goldstone, CA, Madrid, Europe, and Canberra, Australia, which continuously monitor ongoing missions in real time. The Deep Space Network Tracking System provides a two-way communication between Earth-based equipment and the probes, processing data to determine the position and velocity of them.

It is impossible to overstate the importance of the DSN in today's space exploration programme, as Chapter 9 briefly illustrates. "The Deep Space Network[26] is much more than a collection of big antennas. It is a powerful system for commanding, tracking and monitoring the health and safety of spacecraft at many distant planetary locales. The DSN also enables powerful science investigations that probe the nature of asteroids and the interiors of planets and moons" (JPL-NASA 2018, online) (Fig. 2.3).

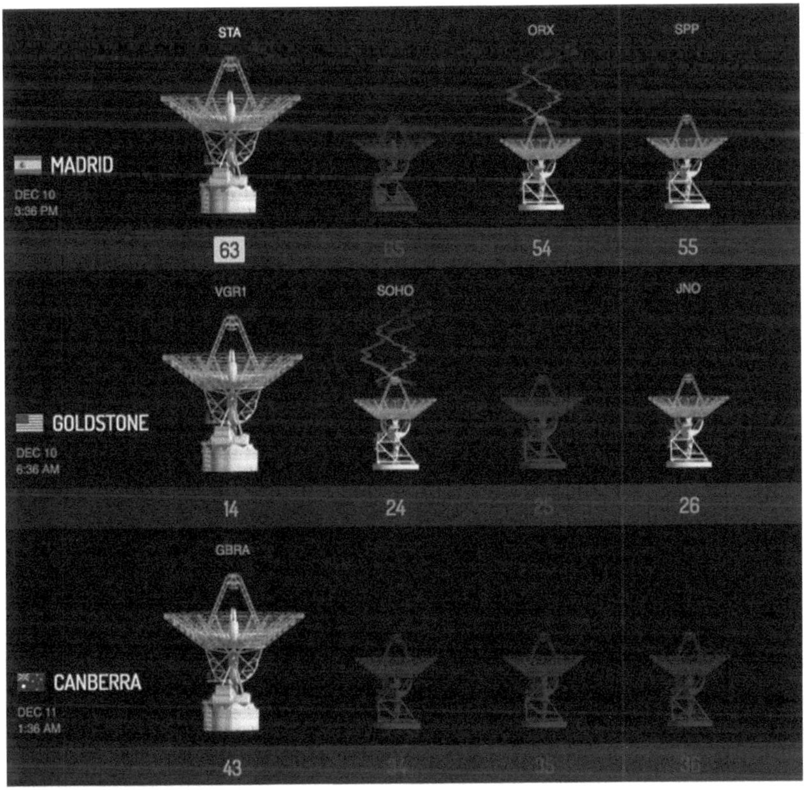

Fig. 2.3 The deep space network in action, December 2018 (This research has made use of the NASA Exoplanet Archive, which is operated by the California Institute of Technology, under contract with the National Aeronautics and Space Administration under the Exoplanet Exploration Program)

In some cases, cooperation happens in officially established multinational organisations. UN-based agencies have been discussed in Chapters 6 and 7, together with the international legal framework of the space sector.

A few do exist at a regional level, though, such as APSCO,[27] an Asian-Pacific organisation China made a lot of efforts to set up in 2008, and whose HQ is in Beijing. Others are, instead, informal bodies with the aim of promoting and supporting international cooperation in space exploration and scientific advancement in space science.

A few examples are the International Space Exploration Coordination Group (ISECG), which is studying the Moon and Mars exploration through international cooperation, and the Intergovernmental Group on Earth Observations (IGEO), which pulls together and analyses Earth observation data to tackle global issues.

In many occasions, however, cooperation takes place on a project basis, as in the case of many European agencies. They, especially ESA, have always been a prominent force for integration, due to more flexible (compared to other countries) laws regulating the sector (see Chapter 7).

The EU[28] as a whole is an important partner in many international initiatives.

Copernicus is a good example. One of the flagship programmes of the EU, Copernicus is managed by the European Commission but implemented in partnership with the member states and other European bodies such as ESA, EUMETSAT, ECMWF and Mercator Océan. Copernicus constitutes the European contribution to the Global Earth Observation System of Systems (GEOSS[29]) and the Committee on Earth Observation Satellites (CEOS[30]).

The peculiar case of Luxembourg has been discussed in its legal aspect on that occasion, but it is the case here to observe that the European Grand-Duchy has put together a space programme through the Luxinnovation initiative and the Luxembourg Space Cluster (Luxembourg Space Cluster 2017), which also involves private international partners, such as the US-based Planetary Resources.

With all this flurry of activities, it is not surprising that the number of international agreements addressing one or more aspects of space activities has kept growing over time.[31] First of all, in terms of number of cooperation activities, which has been growing dramatically since the 1990s, also due to the changing world geopolitics with the fall of the USSR and the opening up of a new era in terms of collaborative efforts among space powers.

But there is more, as Peter explains, and it is due to the fact that "while space science is an established area of cooperation among spacefaring countries, space applications and especially Earth observation are witnessing the emergence of new actors especially in Asia. Fourth, while NASA, as the first space civilian space agency, undoubtedly influenced the institutional structure of subsequent agencies, in the past decade the model of regional space agency inspired by ESA's example seems to have been receiving greater consideration as a new alternative to meet the challenges linked with space activities" (Peter 2016: 15). This point has been clearly illustrated by the recent developments in the Asia-Pacific and in the Americas.

Not everything happens in the same fashion everywhere.

There are countries among which cooperation at an institutional level is easier than in others, especially when the states have previously been partners in defence agreements and military matters.[32] Japan and the USA represent here a good case study, with ongoing cooperation that was first and foremost a comprehensive security alliance and only later on was extended to space missions. As a matter of fact, a 2003 report from the Centre for Strategic and International Studies showed that if the USA helped Japan with a more economically efficient launch vehicle, Japan could in turn develop more valuable military space programs, thus contributing to the regional security. What's more "the US could also reduce overall space data analysis expense by training less experienced Japanese analysts and share the US analysts' workload. Japan's own regional navigational satellite system, the QZSS, could supplement US Global Positioning System" (*The Space Review* 2018, online).

In other cases, cooperation went beyond the typical space club countries; China has recently been one of the countries involved in outreach initiatives, as the already mentioned APSCO,[33] the talks with ESA, and the space-related activities in the wider framework of OBOR/BRI, which has been the object of the case study in Chapter 5.

Institutional cooperation at a global level is mirrored by a number of international joint ventures in the private sector, which have become increasingly common since the end of the 1990s, especially in the launching industry, due to concentration and economies of scale.

Under this regard, the Russians have led the way (Bzhilianskaya 1997). This was to be expected; after the collapse of the Soviet Union, many space actors—including the national space agency itself, which has gone through extensive restructuring—had to shop abroad to continue their space endeavours, engaging with a variety of partners public and private.[34]

Enhanced international cooperation is therefore one of the most promising scenarios in terms of the future of space, as discussed in the conclusions of this book.

Notes

1. The sources used for this chapter are specified in the following notes, but they're generally taken from official reports, national statistics and other verified sources. The problem here is that they're not necessarily comparable from one year to another, due to the inclusion of new segments, as it can be expected from a sector that evolves so fast. Second, countries do not use the same standards, and this accounts for some inconsistencies from one source to another. Moreover, many figures are not disclosed because covered by military secret (this is still common in some countries, such as China, just to mention one). Finally, there has recently been a shift in the literature from the traditional distinction from upstream/downstream operations to a value-chain approach and intangible benefits (which have been examined in this chapter) that somehow complicates things further. As a result, it is fair to say that the sector is at best underestimated, even though the figures provided here are a good indication when taken as a trend.
2. It has to be noted that at times one activity can be classified under one or another category depending on the way it is considered or used. A good example is the navigation systems like GPS, originally planned for military uses but where now the civilian uses are increasingly predominant.
3. This has also an impact on the way space is perceived by the civil society, the level of public support and the overall resources committed to it. This has been analysed in Chapter 8 when talking about the so-called Citizen Space.
4. In 2015, there were two leading measures of the global space economy: the OECD's Space Economy at a Glance (2014) and The Space Foundation's The Space Report (2015). The Space Foundation produces an annual report, referenced here, while OECD has only produced partial updates since then.

5. The growth has been constant in the last five years: the space sector was estimated at just about US$323 billion in 2015 and US$330 billion in 2016.
6. A much-quoted study in the pressed, made by Morgan Stanley, gives an estimate of about US$1.1 trillion space economy in the 2040s (*Space News*, 5 July 2018).
7. Since, at the time of writing, 2018 was still ongoing, year-to-year comparisons have used 2017. However, 2018 partial statistics have been considered in the comments in order to give the right perspective. Which countries feature prominently in launches and an analysis of the upstream segment of the space industry is offered in Sect. 3.3.
8. The first civilian Earth observation system was launched in 1972 by NASA, and it was the Earth Resources Technology Satellite, ERTS-1, later renamed Landsat. See: EOSPSO- NASA's Earth Observing System Project Science Office, official website https://eospso.nasa.gov/.
9. Some claims, however, that this counted towards the Soviet programme, considering that Remek launched from Soviet soil on Soviet carriers. In this case, for the third man in space the world needed to wait until 2003, when China sent Yang Liwei, 2003 on a Chinese spacecraft (Shenzhou) lifted by a Chinese rocket (Long March 2 F). For a detailed discussion about these early stages of space development, see Norberg (2013) and Bednar (2018).
10. "On January 29, 1998, marked an important milestone for the International Space Station as senior government officials from 15 countries met in Washington and signed agreements establishing the framework for cooperation among the partners on the design, development, operation and utilisation of the Space Station" (NASA-ISS 2019, online).
11. It was mission STS-135, which closed 30 years of activity. For the details of the single missions, see the official NASA Space Shuttle's page, at https://www.nasa.gov/mission_pages/shuttle/main/index.html.
12. Dragon was the first private spaceship to dock to the ISS in May 2012.
13. Peter divides the evolution of space technology development in a country can usually be divided into four stages. "The first stage consists of purchasing satellites systems from other countries, the second stage consists of developing such systems in cooperation, the third stage consists of developing the satellite systems independently and the fourth of disseminating knowledge of satellite development to other countries. New perspectives for cooperation will therefore arise among spacefaring countries and newcomers at the interconnection of phases two and four,

reinforcing the trend of multiplication and diversification of relations in space" (Peter 2016: 3).

14. Another traditional way to classify space activities is between upstream (manufacturing and launching spacecraft and satellites into space) and downstream (providing services to the end users), even though a value-chain approach is more widely used.
15. The first commercial trials in suborbital space travel to fly beyond the Kármán line were conducted in 2011 by Virgin Atlantic and 2012 by SpaceX.
16. According to company data, VG in 2014 sold 700 tickets worth $250,000 each (Virgin Galactic 2019, online).
17. In addition to Deep Space Industries and Planetary Resources, other investors, both private and public, are Moon Express (US); ispace (Japan); Asteroid Mining Corporation (UK); Shackleton Energy Company (SEC, US); Kleos Space (Luxembourg); TransAstra (US); OffWorld (US); SpaceFab.US (US); National Aeronautics and Space Administration (NASA, US); European Space Agency (ESA, France); Japan Aerospace Exploration Agency (JAXA, Japan); China National Space Administration (CNSA, China); and Russian Federal Space Agency (ROSCOSMOS). The space mining regulatory framework has been discussed in Chapter 7, while business and technical considerations are included in Sect. 8.2
18. Albeit some of the activities in its long history might come as a surprise. See Chapter 5 for more details.
19. "Within INTA, space technologies currently attract 46% of the total budget. Space capabilities are also strengthened by research in other INTA directorates, including aeronautics, hydrodynamics, security and defence.[…] INTA has a history of strong international collaboration, particularly with NASA and ESA" (INTA 2018, online).
20. Including in the ex-URSS, where not just Russia but also other new CSI countries, such as Kazakhstan and Ukraine, set up one. Their overall number varies according to the source and classification criteria (i.e. only public and/or government funded; institutions in partnership with the private sectors). Physics.org lists as many as 200 (listing also operational centres and not just institutions: https://phys.org/news/2012-06-awesome-space-agencies-world.html) others give more conservative numbers, in between 32 and 85 (WMO-OSCAR database has 85 records) https://www.wmo-sat.info/oscar/spaceagencies.

21. It doesn't mean that Canada has not maintained certain special competences in the launching sector. For example, the Canadian government has funded since the 1970s the development of Canadarm, a robotic arm moving cargo in and out of the cargo bay first of NASA's Space Shuttle. After that, Canada became a world leader in space robotic, also supplying a robotic Mobile Servicing System to the ISS (Bryce Space and Technology 2017). A brief discussion of the Canadian space activities is presented in Chapter 5, Sect. 5.3.
22. Generally, multi-annual budget allocations guarantee a good balance between short-term objectives and long-term planning and avoid the influence of the often-shifting political climate on those deliberations.
23. There are more than the present table suggests, some of them—such as Malaysia, Egypt, Pakistan and Saudi Arabia—with some interesting features. The selection has been done based on relevance to the present book's covered topics.
24. The US space budget is, and has always been, the highest, currently about more than twice the space budgets of the others. Another interesting measure is budget by capita. According to OECD data, Luxembourg spent in 2017 about US$387 per capita (the highest world allocation), USA US$60, UAE US$16.18 and Canada USD 13.5. But all these ratios have to be taken together to convey a fair picture of a country involvement in the sector. India only spends US$0.84%, but the percentage of its budget that goes to the space sector is 0.45%, just slightly less than the USA.
25. Until this July, Australia was the only OECD country without a space agency after the ASO; the Australian Space Office (ASO) had been disbanded in 1996. ASA website: https://www.space.gov.au/.
26. A real-time consultation of DNS currently monitored missions (such as OSIRIS-REX, Juno and even Voyager1) can be checked out at any moment here: https://eyes.nasa.gov/dsn/dsn.html
27. APSCO has been discussed in a case study related to China's activities in Chapter 5.
28. It is impossible to underestimate the importance of international agreements in the space sector, as the lack of them can lead to misunderstanding. An example is given by the launching sector. The EU has not signed any international commercial launch services agreements like the ones the USA has negotiated since the 1990s with a few operators. This has led to some curious case, as when, in 1990, Arianespace accused GWIC of dumping its launch services on the global market. The company

substantiated its allegation with data coming from its US counterparts, which, instead, used the USA–China agreement of 1989 to denounce the violation.

29. "Global Earth Observation System of Systems (GEOSS) is addressing key areas of critical importance to people and society. It aims to empower the international community to: Promote sustainable agriculture; conserve biodiversity; respond to climate change and its impacts; protect populations against natural and human-induced disasters; manage ecosystems and energy resources; understand the environmental sources of health hazards; safeguard water resources; improve weather forecasts" (EU-Copernicus 2018, online).
30. "The Committee on Earth Observation Satellites (CEOS) is an international cooperation mechanism between space agencies and institutions involved in Earth observation. It is responsible for coordinating international civil space-borne missions designed to observe and study planet Earth. CEOS is recognised as the major international forum for the coordination of Earth observation satellite programmes and for interaction of these programmes with users of satellite data worldwide" (Eu-Copernicus 2018, online).
31. Just to have an idea of their numbers and typologies, which include both agreements signed by countries and by the various space agencies, it is worth looking at what NASA has signed over time under its Space Act Agreements (SAAs). In them, NASA enters into SAAs with various partners, domestic and international, over a range of missions and programmes. Their lists and details are regularly updated on the Agency website at https://www.nasa.gov/partnerships/about.html.
32. This is due to the dual-use nature of the space sector equipment and of the existing regulation to their exports, such as the US ITAR (International Trafficking in Arms Regulations). The US Congress modified them in 2013, taking a substantial share of satellites and satellite technologies out of the restricted list to allow their exports to selected countries. "The National Defense Authorization Act (NDAA) for Fiscal Year 2013 repeals a 1999 law that placed all space-related hardware and services, regardless of sophistication or availability, on the U.S. Munitions List, a registry of sensitive technologies whose exports are tightly controlled by the U.S. Department of State. The new law gives the president the discretion to place these items—subject to appropriate national security reviews—on the Commerce Control List (CCL), which contains dual-use items whose exports are regulated by the more business-friendly Department

of Commerce" (*Space News* 2013, online). This list, however, does not include China, to which ITAR still applies. However, China gets access to a few of these technologies through foreign partners, since many European countries have enacted laws promoting "ITAR-free" exports dedicated to the Chinese market.

33. Still, APSCO does represent an important venue for China, due to the fact that the other member states are all less advanced than China in terms of space programmes (Xinhua, 2008). APSCO "allows China to position itself as a purveyor of space technology and expertise to less-developed states. China's leaders also use Beijing's central role in APSCO to promote the export of its space technology and services in order to gain support for its space goals from the Asia Pacific region, as well as to obtain supplementary data and geographic coverage for its space situational awareness efforts" (House of Congress 2016: 35).
34. Some of the most important have been highlighted in Chapter 4.

References

Andsell, M., Lopez, L. D., & Hendrikson, D. (2011). *Analyzing the Development Paths of Emerging Space Nations.* Washington, DC: George Washington University.

Bednar, D. (2018, March 26). *The Geography of Space Exploration: Who Was the Third Country in Space? Medium.* Available at https://medium.com/@danny.bednar/the-geography-of-space-exploration-who-was-the-third-country-in-space-4b7a93ab1f33. Retrieved on 18 December 2018.

Brennan, L., & Vecchi, A. (2011). *The Business of Space: The Next Frontier of International Competition.* London: Palgrave Macmillan.

Broniatowski, D. A., Faith, G. R., & Sabathier, V. G. (2006). *The Case for Managed International Cooperation in Space Exploration.* CSIS. Available at http://web.mit.edu/adamross/www/BRONIATOWSKI_ISU07.pdf. Retrieved on 30 November 2018.

Bryce Space and Technology. (2017). *Global Space Strategies and Best Practices Paper.* Research Paper for Australian Government, Department of Industry, Innovation and Science by Bryce Space and Technology, LLC. s.l.

Bulkeley, R., & Spinardi, G. (1986). *Space Weapons.* NYC: Barnes & Noble.

Bzhilianskaya, L. (1997). Russian Launch Vehicles on the World Market: A Case-Study of International Joint Ventures. *Space Policy, 23,* 329–331.
CBO. (1994). *Reinventing NASA.* Available at https://www.cbo.gov/sites/default/files/cbofiles/ftpdocs/48xx/doc4893/doc20.pdf. Retrieved on 2 January 2018.
Christensen, C. B., Armstrong, K., & Perrino, R. (2016). *Start-Up Space: Rising Investment in Commercial Space Ventures.* AIAA SPACE, Long Beach, CA. Available at https://arc.aiaa.org/doi/pdf/10.2514/6.2016-5233. Retrieved on 30 November 2018.
Cohendet, P. (1997). Evaluating the Industrial Indirect Effects of Technology Programmes: The Case of The European Space Agency (Esa) Programmes. *Proceedings of the OECD Conference on Policy Evaluation in Innovation and Technology.*
Correll, R. R., & Peter, N. (2005). Odyssey: Principles for Enduring Space Exploration. *Space Policy, 21*(4), 251–258.
Dunbar, B. (2017). *What's Next for NASA?* Available at https://www.nasa.gov/about/whats_next.html. Retrieved on 8 February 2018.
EOSPSO-NASA. (2019). *NASA's Earth Observing System Project Science Office Home Page.* Available at https://eospso.nasa.gov/. Retrieved on 25 January 2019.
EU Copernicus. (2018). *Copernicus Homepage.* Available at https://www.copernicus.eu/en. Retrieved on 28 February 2018.
Extance, A. (2014). UAE Fires Up Space Agency with Mars Mission. *Physics World, 27*(11), 11.
FAA. (2018). *The Annual Compendium of Commercial Space Transportation: 2018.* Available at https://www.faa.gov/about/office_org/headquarters_offices/ast/media/2018_ast_compendium.pdf. Retrieved on 11 December 2018.
FAI. (2016). 7 September 1956: The First Human Flies Above 100.000 Feet. *News Page.* Available at https://www.fai.org/news/7-september-1956-first-human-flies-above-100000-feet?type=term&id=1443. Retrieved on 21 September 2018.
Genta, G. (2014). Private Space Exploration: A New Way for Starting a Space-faring Society. *Acta Astronautica, 104,* 480–486.
Haynes, R. (1987). *How We Get Pictures from Space* (Rev. ed.). Washington, DC: U.S. Government Printing Office. Indian Space Research Organisation, 2017. Vision and Mission Statements. Available at https://www.isro.gov.in/about-isro/vision-and-mission-statements. Retrieved 10 February 2018.
Hays, P., Smith, J. M., Van Tassel, A. R., & Walsh, G. M. (2000). Spacepower for a New Millennium: Examining Current U.S. Capabilities and Policies.

In *Spacepower for a New Millennium—Space and U.S. National Security* (pp. 1–35). New York: McGraw-Hill.
House of Congress. (2016, September 27). *Testimony Before the House Space Science and Technology Committee, Subcommittee on Space Hearing on "Are We Losing the Space Race to China?"* Dennis C. Shea Chairman, U.S.-China Economic and Security Review Commission. Available at https://docs.house.gov/meetings/SY/SY16/20160927/105387/HHRG-114-SY16-Wstate-SheaD-20160927.pdf. Retrieved on 18 December 2018.
INTA. (2018). *INTA*. Available at http://www.inta.es/opencms/export/sites/default/INTA/es/. Retrieved on 11 February 2018.
ISU. (2017). *A Roadmap for Emerging Space States*. Strasbourg: International Space University.
JPL-NASA Deep Space. (2018). *Deep Space Network—NASA Jet Propulsion Laboratory Homepage*. Available at https://deepspace.jpl.nasa.gov/. Retrieved on 27 March 2018.
Logsdon, J. (2005). Which Direction in Space? *Space Policy, 21*, 85–88.
Luxembourg Space Cluster. (2017). *Luxembourg Space Capabilities—Luxembourg Cluster Catalogue 2017*. Available at http://clustermembers.luxinnovation.lu/space/wp-content/uploads/sites/4/2016/10/07697_LUXINNOVATION_SpaceCapabilities_05-2017-Web.pdf. Retrieved on 10 January 2017.
Machay, M., & Hajko, V. (2015). Transatlantic Space Cooperation: An Empirical Evidence. *Space Policy, 32*, 37–43.
Mosteshar, S. (2013). The Establishment of the UK Space Agency. In P. Hulsroj, S. Pagkratis, & B. Baranes (Eds.), *Yearbook on Space Policy 2010/2011* (pp. 115–127). Vienna: Springer.
NASA. (2016). *NASA Cargo Launches to Space Station Aboard SpaceX Resupply Mission*. Available at https://www.nasa.gov/press-release/nasa-cargo-launches-to-space-station-aboard-spacex-resupply-mission. Retrieved on 27 February 2019.
NASA. (2019). *NASA—Partners Sign ISS Agreements*. Available at https://www.nasa.gov/mission_pages/station/structure/elements/partners_agreement.html. Retrieved on 27 February 2019.
Norberg, C. (2013). *Spaceflight and Exploration*. Berlin: Springer.
OECD. (2007). *The Space Economy at a Glance*. Paris: OECD.
OECD. (2014). *The Space Economy at a Glance*. Paris: OECD.
Paikowsky, D. (2017). *The Power of the Space Club*. Cambridge: Cambridge University Press.

Pearlman, R. Z. (2011). *NASA's Space Shuttle Program Officially Ends After Final Celebration.* Available at https://www.space.com/12804-nasa-space-shuttle-program-officially-ends.html. Retrieved on 11 February 2018.

Peter, N. (2016). The Changing Geopolitics of Space Activities. *Space Policy, 37,* 145–153.

SIA. (2017). *Satellite Industry Association's 2016 Report.* Available at https://www.sia.org/annual-state-of-the-satellite-industry-reports/2017-sia-state-of-satellite-industry-report/. Retrieved on 27 February 2019.

SIA. (2018). *Satellite Industry Association's 2017 Report.* Available at https://www.sia.org/2018_ssir/. Retrieved on 27 February 2019.

Spaceflight 101. (2017). *Ariane 5 Closes 2017 with Four-Satellite Delivery for European Galileo Navigation System—Spaceflight 101.* Available at http://spaceflight101.com/ariane-5-va240-launches-galileo-m7/. Retrieved on 27 February 2019.

The Space Foundation. (2015). *The Space Report 2015 PDF Is Now Available.* Available at https://www.spacefoundation.org/news/space-report-2015-pdf-now-available. Retrieved on 27 February 2019.

The Space Foundation. (2018). *Space Foundation Report Reveals Global Space Economy at $383.5 Billion in 2017.* Available at https://www.spacefoundation.org/news/space-foundation-report-reveals-global-space-economy-3835-billion-2017. Retrieved on 27 February 2019.

Space News. (2013, January 4). *New Export Law Seen as Boon to U.S. Satellite, Component Makers.* Available at https://spacenews.com/33047new-export-law-seen-as-boon-to-us-satellite-component-makers/. Retrieved on 25 January 2018.

Space News. (2018, July 3). *A Trillion-Dollar Space Industry Will Require New Markets.* Available at https://spacenews.com/a-trillion-dollar-space-industry-will-require-new-markets/. Retrieved on 25 January 2019.

The Space Review. (2018, January 11). *How to Reduce US Space Expenses Through Competitive and Cooperative Approaches.* Available at http://www.thespacereview.com/article/3412/1. Retrieved on 10 December 2018.

Tronchetti, F. (2013). The PCA Rules for Dispute Settlement in Outer Space: A Significant Step Forward. *Space Policy, 29,* 181–189.

UKSA. (2018). *UK Space Agency.* Available at https://www.gov.uk/government/organisations/uk-space-agency. Retrieved on 11 February 2018.

Verger, F., Sourbes-Verger, I., & Ghirardi, R. (2003). *The Cambridge Encyclopedia of Space: Missions, Applications and Exploration.* Cambridge: Cambridge University Press.

Virgin Galactic. (2019). *Virgin Galactic Home Page*. Available at https://www.virgingalactic.com/. Retrieved on 27 February 2019.

Weeden, B. C. (2017). National Space Policy and Administration. In C. D. Johnson (Ed.), *Handbook for New Actors in Space* (pp. 58–60). Secure World Foundation: s.l.

Wood, D., & Weigel, A. (2014). Architectures of Small Satellite Programs in Developing Countries. *Acta Astronautica, 97*, 109–121.

Xinhua. (2008, December 16). *Asia Pacific Space Cooperation Organization Starts Operation*. Available at http://news.xinhuanet.com/english/2008-12/16/conAQ318034tent_10514901.htm. Retrieved on 20 December 2018.

3

The Business of Space

3.1 The Business of Space: General Considerations

The following quotation from the EU's policy paper defines the peculiarities of the space industry, together with its key characteristics: "The space manufacturing industry (satellites, launchers and ground segment) is a strategic, high-tech, high-risk and investment intensive industry with long development cycles and low production rate. In all space-faring nations, space industry mainly depends on institutional programmes, which take two forms: financing of research and development programmes and buying space products and services, as customers of the industry" (European Commission 2013: 3).

What matters the most here is that the space sector is an industry like many others—especially those that manufacture and handle sensitive equipment can be also used for military purposes. Therefore, it should be possible to analyse it with the usual tools of business strategy,[1] business operations and project managements commonly in use. The present section provides a brief introduction to the aspects of the space sector[2] relevant under a business point of view and offers, in this section,

S. Paladini, *The New Frontiers of Space*,
https://doi.org/10.1007/978-3-030-19941-8_3

analytical tools. Some are uniquely conceived for this particular industry, while others come from business studies as mentioned above.

The space sector is characterised by "high risk, high cost, and long payback periods" architecture (Vedda 2009: 99). This has never changed in time, and this is one of the reasons why the space sector has historically needed public support for its development.

It is still the case today, for some specific areas, such as space exploration, which remains, visionary private plans apart (see Chapter 9), a government prerogative. Even in this case, projects have often to compete to attract funding, which can be abruptly cancelled, as the politicians who vote the budgets are their preferences skewed towards short-term results Neal et al. (2011).

If an analysis of the space sector is carried out considering the basic economic concepts of supply and demand, cost versus price, and elasticity of the demand, points of specialty immediately appear and so do a few issues that characterise the sector.

First of all, the pricing policy: "Understanding the distinction between the cost and price of a product or a service isn't always easy in the space sector, since there is a strong heritage of cost-plus contracts. This type of contracts entitles the contractor to a total reimbursement of all the projects costs plus a certain amount of profit, generally based on a percentage of the cost base. It is important to note that the most 'commercial' of the space industries, namely the commercial communications satellite industry, was among the first to migrate away from cost plus contracts" (Gurtuna 2013: 12). Moving to a more sounding approach business-wise in the sector is essential to reduce the high costs the sector is known about and that can doom venture capital companies investing into it.

The literature has identified a set of additional features making the sector different from the others, namely:

- its cyclical nature, which is made worse by the long lead times that characterises it, as it happens in the mining sector;
- its links with the military-defence complex, which explain why so many of the space industry's outputs are subjected to strict regulations that make trade a document-heavy process (see Chapter 6);

- the presence of one prevalent customer—the government—a factor that affects how the business is conducted, in a positive (the state bears the additional costs and shores up the risks) and negative ways (it reduces the willingness to engage in projects not enjoying government support). Not all the countries are prone to this problem in the same way—for example, Germany more than the USA—but it is an issue presents everywhere;
- a static market, with only few operators, which resulted in limited competition and therefore limited innovation. This has started to change in the last two decades, and Chapter 8 is a good starting point to learn about the novelties of the sector;
- a prolonged horizon of investment, since a number of years elapse since the conception of a mission until its realisation and launch. This is not only due to the technical complexity of the spacecraft or the mission, but also to geographical and planetary constraints (favourable launch windows for weather, planetary transits, and so on).

A telling example is the ESA-Rosetta case, a well-known success story in space exploration. Not many people are aware that mission was originally intended for the Comet 46P/Wirtanen. Approved in 1993, it was to begin in 2003, but technical problems delayed the launch. As a result, another celestial object, Comet 67P/Churyumov–Gerasimenko, was selected instead, and Rosetta was launched in 2004 to reach its destination in 2014. The study lasted 18 months. "Taking into account the time required to analyse the result from the encounter with the comet, the project will have an overall life-span of around 25 years" (Noberg 2013: 1).

When this happens with products that target consumer needs, however, as in the case of satellite telecommunication, this delay can become extremely dangerous for the providers, as those needs and those conditions may well change before the products are ready.[3]

- the lack of economy of scale in terms of production, which, of course, makes the costs skyrocket fast. More in general, cost is still a real entry barrier to the space industry and, without addressing the launching industry's constraints, this situation will not get better. In addition to that, the costing exercise[4] itself presents specific difficulties in space projects,[5] which can explain why many of them ended up costing more than what originally budgeted;

- linked to the former point, there is the issue of limited access to the financial markets. Even after the sector opened to private investment, and with the exception of communication satellites, space projects are in the majority of the cases carried out with government funding or, at best, in PPP. Until a more efficient system of raising capitals is devised, these specific challenges will stay.
- risk management[6] in space has a different profile than on Earth, both in terms of definition of what constitutes a risk and in terms of modelling and mitigating its occurrence.

Nowhere as in the upstream segment, for example—the launching industry, are the peculiarities of the space sector more evident.

Launching is, traditionally, a crucial segment of the whole sector: in any space mission, the main challenge is constituted by the huge costs of reaching Earth's orbit. This is a constraint that applies to virtually all the operations, from manned flights to satellites and to space exploration by probes or by space telescopes.

Until recently, those costs have remained stable.

Even with SpaceX's comparatively more efficient approach, as of 2018, the average cost of launching cargo into orbit remains equivalent to the times of the Space Shuttle, e.g. about $30,000 per lbs (Business Insider 2018). What happened is that, since the demand has remained constant-due to the long lead times in building satellites and probes—the result was an oversupply in terms of launching capabilities and pressure on price reduction.

There is also another issue that affects the launching industry, and it is a destination problem: apart from space exploration—which, as shown in Chapter 9, is still the business of a handful of states—there is no demand for space destinations on a regular basis. The only exception is the ISS, served at the moment by Soyuz for the astronauts and by a few private operators for cargoes. Until space tourism or, even farther in time, out-of-Earth space stations and colonies develop, this fact will represent a bottleneck to the development of a competitive launching industry.

Finally, in a business analysis of the space sector, a question should be asked about the benefits that space programmes bestow to the investors, both countries and privates. And while direct benefits are evident—there is an abundant literature[7] about all the technical applications in everyday use that have come from space projects—there are other, less evident returns, most of them intangible, which accrue on the long run.

3.2 The Star of the Sector: The Satellite Industry

As mentioned in Chapter 2, the satellite market represents the mature segment in the space sector, in continuous expansion since the early days of the space age.

But how many satellites are there in Earth's orbit at the moment? This is a question worth asking because, as mentioned in Sect. 2.1, since the 1970s where the civilian satellite programmes have started, this number has increasingly grown in absolute and relative ratio and in terms of typology of satellites in use.

The United Nations Office for Outer Space Affairs (UNOOSA) is the international agency with the mandate to monitor this staggering number (whose inherent risks have been considered in detail in Chapter 6), which will be also discussed in Chapter 7 when talking about crowded orbits and their risks. Its registry, called the Index of Objects Launched into Outer Space,[8] is periodically updated upon new launches and retirement of out-of-service satellites (Fig. 3.1).

The breakdown of the existing satellites and probes into spaces by country/organisation since 1957 has been given in Chapter 7. For now, it is enough to say that the number of countries involved in the space sector has increased over the years to more than sixty. As of the end of 2018, there were 4919 satellites orbiting planet Earth, an increase of 4.79% compared to 2017 in absolute number. In terms of objects launched each year, in the last 11 months (January–November), UNOOSA has incremented the registry with 204 objects launched into space. To give an idea of the numbers, this is more than the yearly average before 2013, and about the average of 2013, 2014, 2105 and 2016.

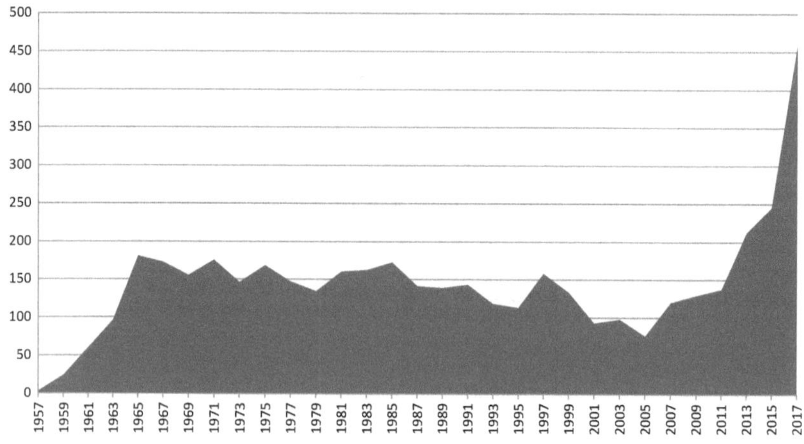

Fig. 3.1 Cumulative number of satellites sent into space by year (1957–2017)

The 2018 statistics, however, are lower than the rather exceptional number of 2017, with 453 satellites launched, about 5% of the objects ever sent to space. According to FAA,[9] the total of 2017 of object sent to orbit was even higher—469 (FAA 2018). 23 of them failed to reach their destination (due to launch failures). In addition to satellites, a remaining 3% were crew-supporting vehicles and cargo to the ISS (14); one was an X-37B mission, and one was a telemetry test package carried by Rocket Lab's maiden launch.

While it is true that a significant part of them were micro-objects (the so-called CubeSats[10] for Earth observation and/ communication purposes), the overall number is still impressive.

SIA (2018) reports a lower number—still a record—with 345 satellites launched in 2017, leading to an estimated total of 1901 active satellites orbiting the Earth[11] in December 2017 (an increase compared to 1459 active satellites in December 2016).

An interesting snapshot is a more in-depth analysis of the satellites currently in activity in terms of their use, again coming from the SIA reports (2017, 2018).

With 1901 satellites currently in operation, the growth since 2012 has been nothing but impressive (91% over six years, from 994 satellites in operation in 2012), mainly due to small and very small satellites in LEO

Table 3.1 Global satellite industry revenue 2006–2017

Year	US$ billion	Growth %
2007	122	15
2008	144	19
2009	161	11
2010	168	5
2011	177	6
2012	210	18
2013	231	10
2014	247	7
2015	255	3
2016	261	2
2017	269	3

Source Author's elaboration on SIA data (2018)

(<1200 kg). The satellites launched in between 2012 and 2016 increased 53% over previous 5 years, with an average of 144/year.

But there is another factor responsible for this record number of orbiting satellites: the average operational duration of communications satellites—which are generally the biggest ones—is becoming longer, exceeding 15 years; as a matter of fact, 247 still active satellites have been launched before 2002. The communication satellites make up a big share of the total (35% in 2016), EO coming second with 19%.

Both technology developments and improved interest in space, particularly from start-up companies, are behind the recent drive. A good way to appreciate the growth and the *richesse* of the sector is to go through SIA's yearly reports[12] released by the Association on the state of the market since 1997.

On a space industry worth US$348 billion overall,[13] the satellite industry accounted in 2017 for about 77% of it (Table 3.1).

Table 3.2 shows the revenues of satellite industry since 2006, with figures that have more than doubled since the beginning. The yearly average of the US market share was about 44% of the total (accounting in 2016 for US$110.3 billion over a global US$260.5 billion). However, while in previous years the US growth outstripped the rest of the world, the last three years have seen an opposite trend.

Table 3.2 Global satellite manufacturing revenues in the global space industry (2017)

	US$ billion
Government space budgets	79.3
Satellite services	128.7
Ground equipment	119.8
Satellite manufacturing	15.5
Launch industry	4.6
Total	347.9

Source Author's elaboration on SIA data (2018)

In 2017, the growth in the satellite industry was less impressive compared to the past—where in some cases has reached double-digit increases—but still substantial. The whole segment grew by 3% over 2016 (up to US$268.6 billion), split down into four main subsegments:

(1) Satellite manufacturing: This category includes "launch vehicles and subsystems, satellites/payloads/spacecraft and subsystems, ground segment systems and equipment (control centres and telemetry), suppliers of materials and components, scientific and engineering support, fundamental and applied research" (UKSA 2018: 4). With 10% increase in 2017 (US$15.5 billion, 8.8 billion of which in the US only—about 6% of the total) satellite manufacturing led the year growth, also due to the high number of satellites sent into orbit. The whole subsegment is now worth US$15.5 billion, in counter-trend compared to the last two years, where manufacturing either stagnated or fell by—13.9% year-over-year, as shown in Table 3.3.
Manufacturing of satellites for military surveillance (44% of the total in 2016) and commercial communications (18%) is still the strongest items, even though with relative variations depending on the base year. The statistics, however, only tells a part of the story. As for aircraft construction, tracking orders is another way to gauge the state of the sector. For example, orders for 17 commercial GEO satellites to be delivered in the following years were announced in 2016, ad while down in respect of 2010 (with 28 orders) the number was still significant.

Table 3.3 Time series of commercial launches 1990–2018

	1990	1991	1992	1993	1994	1995	1996	1997	1998	1999	2000	2001	2002	2003	2004	2005	2006	2007	2008	2009	2010	2011	2012	2013	2014
Total space launches	**15**	**12**	**14**	**11**	**15**	**23**	**24**	**38**	**41**	**39**	**35**	**16**	**24**	**17**	**15**	**18**	**21**	**23**	**28**	**24**	**23**	**18**	**20**	**23**	**23**
United States, total	**9**	**6**	**6**	**5**	**5**	**12**	**11**	**17**	**22**	**15**	**7**	**3**	**5**	**5**	**6**	**1**	**2**	**3**	**6**	**4**	**4**	**0**	**2**	**6**	**11**
Antares	0	0	0	0	0	0	0	0	0	0	0	0	0	0	0	0	0	0	0	0	0	0	0	0	3
Athena	0	0	0	0	0	1	0	1	1	3	0	0	0	0	0	0	0	0	0	0	0	0	0	2	0
Atlas	1	2	3	3	4	8	7	6	5	4	3	1	3	4	5	1	1	0	1	1	0	0	0	0	1
Conestoga	0	0	0	0	0	1	0	0	0	0	0	0	0	0	0	0	0	0	0	0	0	0	0	0	0
Delta	5	4	3	1	1	1	3	7	11	5	2	1	2	0	0	0	1	3	2	2	2	0	0	0	1
Dragon	0	0	0	0	0	0	0	0	0	0	0	0	0	0	0	0	0	0	0	0	0	0	0	0	0
Electron	0	0	0	0	0	0	0	0	0	0	0	0	0	0	0	0	0	0	0	0	0	0	0	0	0
Falcon	0	0	0	0	0	0	0	0	0	0	0	0	0	0	0	0	0	0	2	1	2	0	2	3	6
Minotaur I	0	0	0	0	0	0	0	0	0	0	0	0	0	0	0	0	0	0	0	0	0	0	0	1	0
Pegasus	0	0	0	1	0	1	1	3	4	2	2	0	0	1	0	0	0	0	1	0	0	0	0	0	0
Taurus	0	0	0	0	0	0	0	0	1	1	0	1	0	0	1	0	0	0	0	0	0	0	0	0	0
Titan	3	0	0	0	0	0	0	0	0	0	0	0	0	0	0	0	0	0	0	0	0	0	0	0	0
Europe, total	**5**	**6**	**6**	**6**	**8**	**8**	**9**	**11**	**9**	**8**	**12**	**8**	**10**	**4**	**1**	**5**	**5**	**6**	**5**	**5**	**6**	**4**	**6**	**4**	**6**
Ariane 4	5	6	6	6	8	8	9	11	9	8	8	6	7	1	0	0	0	0	0	0	0	0	0	0	0
Ariane 5	0	0	0	0	0	0	0	0	0	0	4	2	3	3	1	5	5	6	5	5	6	4	6	3	4
Soyuz 2	0	0	0	0	0	0	0	0	0	0	0	0	0	0	0	0	0	0	0	0	0	0	0	1	2
Vega	0	0	0	0	0	0	0	0	0	0	0	0	0	0	0	0	0	0	0	0	0	0	0	0	0
Russia, total	**0**	**0**	**0**	**0**	**0**	**0**	**2**	**7**	**5**	**13**	**13**	**3**	**8**	**5**	**5**	**8**	**9**	**12**	**11**	**10**	**13**	**10**	**7**	**12**	**4**
Cosmos	0	0	0	0	0	0	0	0	0	1	2	0	0	1	0	1	0	0	3	0	0	0	0	0	0

(continued)

Table 3.3 (continued)

	1990	1991	1992	1993	1994	1995	1996	1997	1998	1999	2000	2001	2002	2003	2004	2005	2006	2007	2008	2009	2010	2011	2012	2013	2014
Dnepr	0	0	0	0	0	0	0	0	0	1	1	0	1	0	1	0	1	3	2	1	3	1	0	2	2
Kosmos	0	0	0	0	0	0	0	0	0	0	0	0	0	0	0	0	1	2	0	0	0	0	0	0	0
Proton	0	0	0	0	0	0	2	6	4	5	6	2	5	1	4	4	4	4	6	7	8	7	7	7	2
Rockot	0	0	0	0	0	0	0	0	0	0	0	0	2	1	0	1	1	0	0	2	1	0	0	1	0
Shtil	0	0	0	0	0	0	0	0	1	0	0	0	0	0	0	0	0	0	0	0	0	0	0	0	0
Soyuz	0	0	0	0	0	0	0	0	0	6	3	0	0	2	0	1	0	0	0	0	0	0	0	0	0
Soyuz 2	0	0	0	0	0	0	0	0	0	0	0	0	0	0	0	0	1	3	0	0	1	2	0	1	0
Start	0	0	0	0	0	0	0	1	0	0	1	1	0	0	0	0	1	0	0	0	0	0	0	0	0
Volna	0	0	0	0	0	0	0	0	0	0	0	0	0	0	0	1	0	0	0	0	0	0	0	0	0
Zenit 3SLB	0	0	0	0	0	0	0	0	0	0	0	0	0	0	0	0	0	0	0	0	0	0	0	1	0
Ukraine, total	**0**	**0**	**0**	**0**	**0**	**0**	**0**	**0**	**1**	**0**	**0**	**0**	**0**	**0**	**0**	**0**	**0**	**0**	**0**	**0**	**0**	**0**	**0**	**0**	**0**
Zenit 2	0	0	0	0	0	0	0	0	1	0	0	0	0	0	0	0	0	0	0	0	0	0	0	0	0
China, total	**1**	**0**	**2**	**0**	**2**	**3**	**2**	**3**	**4**	**1**	**0**	**0**	**0**	**0**	**0**	**0**	**0**	**0**	**0**	**1**	**0**	**2**	**2**	**0**	**0**
Long March 2C	0	0	0	0	0	0	0	1	4	1	0	0	0	0	0	0	0	0	0	0	0	0	0	0	0
Long March 2E	0	0	2	0	1	3	0	0	0	0	0	0	0	0	0	0	0	0	0	0	0	0	0	0	0
Long March 3	1	0	0	0	1	0	1	0	0	0	0	0	0	0	0	0	0	0	0	0	0	0	0	0	0
Long March 3B	0	0	0	0	0	0	1	2	0	0	0	0	0	0	0	0	0	0	0	1	0	2	1	0	0

(continued)

Table 3.3 (continued)

	1990	1991	1992	1993	1994	1995	1996	1997	1998	1999	2000	2001	2002	2003	2004	2005	2006	2007	2008	2009	2010	2011	2012	2013	2014
Long March 2D	0	0	0	0	0	0	0	0	0	0	0	0	0	0	0	0	0	0	0	0	0	0	1	0	0
India, total	**0**	**0**	**0**	**0**	**0**	**0**	**0**	**0**	**0**	**0**	**0**	**0**	**0**	**0**	**0**	**0**	**0**	**1**	**0**	**0**	**0**	**0**	**0**	**0**	**1**
PSLV	0	0	0	0	0	0	0	0	0	0	0	0	0	0	0	0	0	1	0	0	0	0	0	0	1
Sea Launch[a], total	**0**	**0**	**0**	**0**	**0**	**0**	**0**	**0**	**0**	**2**	**3**	**2**	**1**	**3**	**3**	**4**	**5**	**1**	**6**	**4**	**0**	**2**	**3**	**1**	**1**
Zenit 3SL	0	0	0	0	0	0	0	0	0	2	3	2	1	3	3	4	5	1	6	4	0	2	3	1	1

Key: U = data are not available

[a]Sea Launch is an international venture involving organizations in four countries and uses its own launch facility outside national borders. Their first commercial launch, in 1999, was licensed by the Federal Aviation Administration. Sea Launch filed for Chapter 11 bankruptcy protection in June 2009 and thus had no launches in the second half of the year

Notes A commercial launch is a launch that is internationally competed (i.e., available in principle to international launch providers) or whose primary payload is commercial in nature. FAA-licensed launches carrying captive government (NASA and DOD) or industry payloads (ORBCOMM, Delta 3 demosat, Zenit 3SL demosat, and others) are counted here. Data are for orbital launches only

Source

1990–1999: U.S. Department of Transportation, Federal Aviation Administration, Associate Administrator for Commercial Space Transportation, personal communication, 4 June 2002

2000–2014: U.S. Department of Transportation, Federal Aviation Administration, Commercial Space Transportation: Year in Review (Washington, DC: Annual Issues), Table 2, available at http://www.faa.gov/about/office_org/headquarters_offices/ast/ as of 19 May 2015

2015–2016: U.S. Department of Transportation, Federal Aviation Administration, Office of Commercial Space Transportation, The Annual Compendium of Commercial Space Transportation, Appendix 2, available at https://www.faa.gov/about/office_org/headquarters_offices/ast/new_publications/ as of 27 December 2017

2017: U.S. Department of Transportation, Federal Aviation Administration, Commercial Space Data, Launches, available at http://www.faa.gov/data_research/commercial_space_data/launches/?type=Licensed as of 28 June 2017

2. Launching services: US$4.6 billion, with the non-US representing 61% (US$2.8 billion) of the total. The USA operated the majority of 2017 commercial launches. UKSA labels them as "space operations".
3. Ground equipment: (e.g. satellite navigation equipment, GNSS; traffic information systems; aircraft avionics, maritime, surveying and rails), which in other reports are listed as "space applications[14]": US$119.8 billion, with 6% growth compared to the previous year. Of these, US$108 billion went into consumer equipment, such as GNSS, TV dishes and so on.
4. Satellite services: (such as broadband, communication and Earth observation services), labelled in other reports as "space applications". They remained the leading segment in 2017, accounting for overall US$128.7 billion. The television represents here the lion share (28%) with US$98 billion, followed by fixed communication (US$17.9 billion). There are about 220 million satellite pay-tv subscribers worldwide, and the segment is driven by a growing demand in emerging markets. This is one of the mature segments, however, and there is a potential for a slowdown in demand for satellite capacity, as the compression technologies continue to improve. On the other hand, broadband (US$2.1 billion, 5% growth compared to the previous year), mobile (US$4 billion) and EO (US$2.2 billion) still represented a small part of satellite services, even though there is scope for growth faster than anticipated due to the additional capacity available on newly launched satellites in the USA. Moreover, the number of non-US subscribers is growing.

Not all satellite applications are at the same level of maturity in terms of market.

While communications appear to have reached a "mature" phase, EO (Earth observation also called remote sensing) is still in expansion. Moreover, while EO was one of the earliest uses of satellites, Commercial Observation Satellites (COS) are still relatively new, but applications connected to environmental concerns and climate change—like disaster mitigation, geographical condition surveillance and meteorology—are attracting now a growing interest.

EO as a whole has gone through significant changes in the last ten years. OECD reported in 2007 that the majority of EO services were provided by just a handful of providers (less than ten), while, in the last two years, the sector has seen new entrants, huge investments and the development and deployment of more satellites.

New partnerships have emerged, such as Airbus—EADS Astrium (2013), Planet—BlackBridge (2015), MDA—DigitalGlobe (2017), Planet—Terra Bella (2017) and UrtheCast—Elecnor/Deimos (2015). In this last case, UrtheCast operates cameras aboard ISS and acquired assets from Elecnor Deimos, but is also planning to deploy its own optical and radar satellites.

Other incentives are given by the diffusion of technology transfer programmes—the UK-based Surrey Satellite Technology Limited (SSTL) is a good example—and by the multiplication of microsatellites, such as CU-Sat and other small, cheap operators, as examined in Chapter 8.

If the offer is multiplying, so it is the demand for these services. According to some estimates, 23 countries are working on a dedicated remote sensing system and, by the end of the decade, Asia is going to be in the lead. A few of them (India, China, Malaysia, Thailand, Singapore, Vietnam, Taiwan, South Korea and Japan among them) have all shown interest in this sense.

Finally, the launching industry suffered a decrease in 2017, of about 16% in terms of revenues. This was the only subsegment to register a negative growth, albeit the increment in orbital launches, due to the reduction of overall costs, in one of the ongoing revolution of the sector. The next two sections examine the launching segment in closer details.

3.3 Upstream Space: The Launching Industry

At the beginning at the space era and in the 1960s, there was no possible confusion regarding who was sending probes and rockets into space. There were few countries and their governments behind them, and security concerns were at the origin of those efforts, as the brief history recap in the previous section has shown.

Table 3.4 2016–2018 orbital launches by country

	2015	2016	2017	2018	*2019*[a]
China	19	22	18	39	*6*
Europe	9	9	9	8	*2*
India	5	7	5	7	*2*
Iran	1	–	1		*2*
Israel	–	1	–	–	–
Japan	4	4	7	6	*1*
New Zealand	–	–	1	3	*1*
North Korea	–	1	–	–	–
Russia	29	19	20	20	*5*
Ukraine	–	–	1	–	–
USA	20	22	29	31	*7*
Total	87	85	91	114	*26*

[a]Partial data up to April 2019
Source Author's elaboration on official data from declaring countries; space.skyrocket.de

Things have changed a great deal since then, in terms of size but also in terms of actors. It is true that, when launching is about, space still belongs to the pioneers, those states that first initiated the space race and the ones who have followed up shortly to carve out their own share. This picture has become, however, increasingly nuanced, and while states and governments remain the main players, commercial and private companies now represent an important part in the space industry. If the distinction was in the past between military and civilian (with the civilian generally represented by space agencies), it is nowadays increasingly between public and private.

When retrieving the launching statistics, a first distinction that has to be made is between launches by their source; when the analysis is restricted to commercial launches,[15] which are the ones of specific interest in this book, numbers are different. The commercial share of total launches has varied over the years due to a series of factors, such as the satellites sent to orbit, even though often one carrier delivers more than one. As a comparison, 2010 it was 31% of the total, with 23 launches as a whole in a total of 74.[16]

Table 3.4 provides a time series of almost thirty years of commercial space launches together with the indication of the rockets used for them.

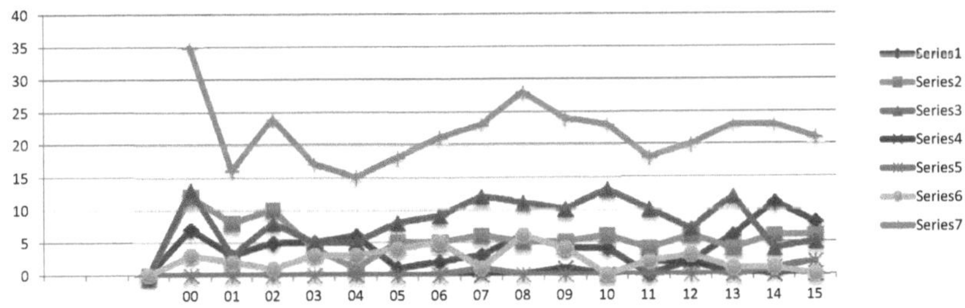

Fig. 3.2 Commercial space launches conducted by selected countries, 2000–2015 (*Source* Author's graph on countries' official statistics, different years)

Other interesting data for commercial launches is their destination in Earth's orbit, not present in the time series but still worth checking. A look at the statistics also shows that the already noticed trend of expansion in the number of operators/countries active in the space industry is mirrored in the number of commercial launches, with a substantial increase over time.

Commercial launches are not a novelty of the present decade.

The first took place in 1989 from the White Sands Missile Range in New Mexico, with a Starfire sub-orbital vehicle that carried a Consort-1 payload—but their importance are increasing. Figure 3.2 offers a comparison of commercial launches conducted by selected countries or regions from 2000 through 2017.

An analysis of the last five years (2015–2019) of orbital launches in terms of countries offers additional insights (Fig. 3.3).

In 2017, the official statistics gave a 7% increase in the number of orbital launch attempts worldwide 90 (or 91) from sites in seven different countries. Of 91 total launches[17] in 2017, 64 commercially procured launches the others were 13 non-commercial and 13 involved space vehicles—were conducted in 2017, the same than in 2016, with the USA with the largest quotas of them (39%). There have been a 100% increase in the total number of spacecraft deployed and a 200% increase in the number of commercial spacecraft deployed.

In 2018, those figures were even more impressive, with 114 launched attempted and only 2 failures[18] (The Space Report 2019). This was the

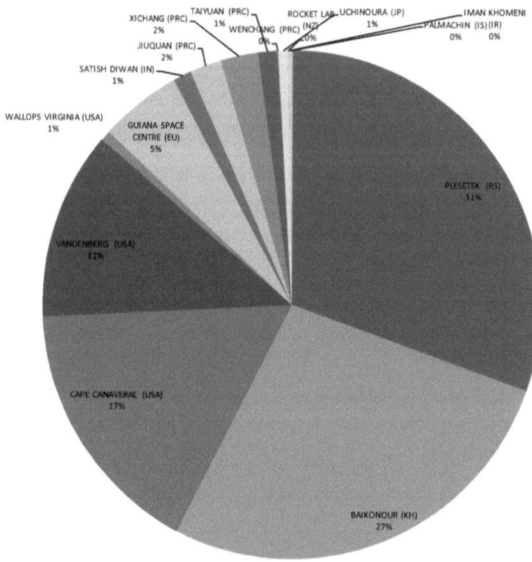

Fig. 3.3 The world's spaceports: selected launching sites (*Source* Author's graph on countries' official statistics, different years)

highest number since 1990, in yet another sign that the growth in the sector is accelerating.

Another interesting breakdown in for destination in terms of Earth's orbit, which shows why space, as infinite as it might seem, is a finite resource after all, and some of the orbits are filling up quite fast: in 2016 of about 64 satellite launches, 28 were into LEO, 28 into GEO, 6 into MEO and only 2 beyond GEO (SIA 2017).

Breaking down the figures by countries, the USA has dominated the launch market in 2017, and the sector remained strong even in 2018 with a 13% growth, albeit having been overcome by China as the world's leader.

In 2017, 23 launches were conducted under licenses issued by the Federal Aviation Administration's Office of Commercial Space Transportation (FAA AST), which represents for the USA the highest number since 1989, when the first commercial launch took place. All except one were orbital. This took the total amount of the US statistics of launches between 1990 and 2017 to 213.

In addition to the stellar performance of SpaceX, Elon Mask's company is alone responsible of about 1/5 of all launches carried out in the year; there are other major operators, like Arianespace and International Launch Services (ILS), which compete with the American providers. The US share in the market—54%—is expected to raise even further as soon as new domestic operators—Virgin Orbit and Rocket Lab—enter the fray.

It has not always been like that; a look at the tables will show that in the former decade, the US commercial launches were fairly limited and in 2011 there was not even one. The decline appears even more substantial if compared with data since the times of the Space Shuttle. "The U.S. Fell from 66 satellite launches in 1998 to zero in 2011. Private space launches are reversing that trend" (*Popular Mechanics* 2013, online). Those days of Russian (and to lesser degree, European) dominance are gone. Since 2014, US providers—especially private—have conquered shares year after year, and the forecast looks positive.

Who launches what can be somehow tricky to assess.

While it is true that "most of this launch activity is captive; that is, the majority of satellite operators have existing agreements with launch service providers or do not otherwise 'shop around' for a launch" (FAA 2018: 4), the commercial satellite launch market is becoming more diversified by the day, with many operators using piggybacking as a secondary payload on a multiple-satellite launch. There are still a few restrictions in place: the US government satellites are still obliged to use US rockets. Moreover, all the US rockets have to launch from the US territory, and the US Sea Launch in the Pacific Ocean has never allowed foreign vectors.

Another part of the sector regards spaceports. For what concerns the USA, commercial launches can lift off either from the two federal launches site (Cape Canaveral, Florida and Vandenberg, California) or one of the eight spaceports shown in the map. The oldest of them is California Spaceport, which received its licence in 1996 (Fig. 3.3).

In 2018, the country saw 34 orbital launches, all successful. Half of them take off from Cape Canaveral Air Force Station in Florida, which was, together with Xichang Satellite Launch Centre in China, the world's first spaceport. Together, the two hosted 30% of the world's launches in 2018 (The Space Foundation 2019).

So far for what concerns the USA.

Russia still holds on as one of the prominent launchers—after all, it is the only country that now brings astronauts to the ISS after the retirement of the Shuttle and in the wait of SpaceX to qualify for manned flights—but its share is shrinking. While carrying out about half of the launches in 2013, now it is down only 21 orbital launches in 2017 and 20 in 2018 (only 19 successful, due to the Soyuz FG's failure).

The country's Soyuz workhorse is still the most used of the Russian rockets, launching 15 times in 2017, four of them to take the astronauts to the ISS; Proton flow 4 missions while Rokot, that approaches its retirement, only once. Among those 21, only three have been commercial launches. 2018 saw similar statistics, with the Soyuz launching 16 times.

If China is the third country, after Russia and USA, in terms of object ever launched into space (see Chapter 5) since the beginning of its space programme, it is now featuring higher in the ranking, having emerged over the years as a provider of launching systems, both government-led and commercial.

The steadily growing number of Chinese orbital launches since 2010 proves it. 2017 saw 18 launches, 10 of them civilians. This was due to the substantial investments poured in the space sector, which translated into new launching vehicles (Kaituozhe 2, Kuaizhou 11, Long March 5, Long March 6, and Launch March 7), a new launch site on Hainan Island, and investments in a new space station.

Data for 2018 were record shattering, with 39 launches, higher than the previously most successful year, 2016 (22). China's share of the global orbital market was at 34% in 2018, overcoming the USA (30% of share) as the world's first.

European vectors—the Arianespace's family of rockets—managed to launch 11 times in 2017 but only 8 in 2018; this share is however due to increase when the new Ariane 6 finishes its testing and starts commercial operations (see Chapter 4).

In addition to these four countries, only India, Japan and New Zealand managed to put payload into orbit in the last two years, even though other countries have launching capabilities or are currently developing them. As such, the launching statistics provide a good snapshot of the global space industry, both in terms of evolution of the sector over time and the main players, which are, increasingly, private spaceflight companies, discussed in the next section.

3.4 New Actors for a Changing Sector: The Rise of the Private Spaceflight Operators

As already mentioned, spaceflight has been the latest segment in the space sector to open up to the private sector as a whole (with the notable exception of Arianespace that, however, has been always a private–public partnership with the involvement of many public stakeholders, the France government and ESA *in primis*).

And, as far as private ventures go, nobody more than SpaceX is in the spotlight, for a good reason: since the moment private venture capital entered the launching sector, nobody has done more than SpaceX. Its Dragon module became in 2010 the first (and only one to date) private company to return a spacecraft from LEO and two years after to deliver the first—but the not last—cargo to the International Space Station. As if that were not enough, the enterprise launched by the visionary Elon Musk has also made a bid to take humans to Mars. And, while SpaceX is not the only private company that developed rocket technology for space missions, it is the only one where the outer space beyond the Earth's Orbit features prominently in the company plans. In comparison, the others—like Virgin Galactic—see themselves restrained into LEO and suborbital flights, even though Blue Origin (owned by Amazon's tycoon Jeff Bezos) has recently declared an interest in reaching out for the Moon.

As a pioneer and an example for all the private sector's newcomers, SpaceX deserves a closer look. The company was founded in 2002 by Tesla owner's Elon Musk and employs as of 2018 more than 6000 people. With three working vehicles (Dragon, Falcon 9, and Falcon Heavy), it had more than 100 contract flights in 16 years of activities and represents alone a giant in the launching sectors. SpaceX carried out 20% (18) of all the launches in 2017, all with Falcon 9,[19] the ordinary vehicle that resupplies the ISS. 2017 was also a failure-free year for SpaceX, which speaks volumes about its reliability. Falcon 9's performance in 2018 was slightly lower in terms of world's share (18%), but not of launches (21).

What does make Elon Musk's company different from the others? A few, but important details. First of all, Dragon is the spacecraft that has replaced the Shuttle for resupplying the ISS, with many missions flown every year. While it still cannot ship astronauts to the space station—the Russian Soyuz[20] had the virtual monopoly since 2011—it has consistently delivered payload and science materials starting from 2012. SpaceX also performs many other functions, including putting into orbit satellites for third countries. A recent example is Bangladesh's first satellite—the Bangabandhu-1—launched on 11 May 2018.

However, where SpaceX has been really innovative and ground-breaking is that it translated the logic behind the Space Shuttle family (NASA and its less-known USSR cousin, Buran), i.e. a vehicle that can be launched over and over to space, to the rocket segment, using not a shuttle but component-made stages.

The reuse of rocket stages (or the lack of it) is one of the nodal points that make space travels so expensive (the other is the amount of propellant needed to lift-off against Earth's gravity, a factor that will be discussed in Chapter 9). Therefore, a reusable launch vehicle (RLV) brings down the cost[21] of orbiting per kg of payload by an order of magnitude, through reuse of the vehicle systems. SpaceX has already reached a few milestones, as in March 2017 with the historic first re-flight of an orbital class rocket (Space.com, 31 March 2017). Since then, the company has regularly launched "flight-proven" rockets; in 2017, it completed 14 first-stage recoveries with 100% of success rate and then flying five missions with those recovered boosters.

In 2018, SpaceX launched Falcon Heavy, the world's most powerful operational rocket and the one that should (one day) take humans to Mars. Its first launch[22] fired off the Kennedy Space Centre in Florida on 6 February, succeeding in its stage and booster separation and booster recovery (the two sides landed as planned at SpaceX Landing Sites 1 and 2), even though the core did not land, as it was planned, on SpaceX's ship on the Atlantic Ocean.

SpaceX's plans ahead are ambitious: Elon Musk's company does not only aim at carrying out cargo deliveries, but also qualify to take astronauts to the ISS and beyond Earth's Orbit. Its manned flight accreditation is ongoing[23] and it is expected to be completed by 2019. Its competitors are like-minded.

This is going to be a sustainable path if, and only if, those enterprises will earn enough money to justify the huge amount of costs and risks associated with the missions. This has already happened in history, during the great sea exploration in the 1500s, which proved so profitable that many others followed Colombo and Magellano in the scramble for new world's riches.

In terms of perspectives, the best possible solution for all the players would be a comprehensive cooperation between the private and public sector to maximise the results, and nothing would harm more the space sector as a whole that a scenario in which the space agencies fight private space operators. A climate of domestic and international cooperation seems essential in this case: "NASA spokesman David Steitz, speaking about the plans of Inspiration Mars[24] for sending a spacecraft around Mars, said it is a testament to the audacity of America's commercial aerospace industry and the adventurous spirit of America's citizen-explorers… NASA will continue discussions with Inspiration Mars to see how the agency might collaborate on mutually beneficial activities that could complement NASA's human spaceflight, space technology and Mars exploration plans" (moonandback, 27 February 2013).

NASA's intentions matter: SpaceX is not alone in this race to space. Other private companies feature in these last years' launching statistics, with United Launch Alliance, Sierra Nevada, Orbital ATK, among them. Not all of them are US or European: China, whose country profile has been covered in Chapter 5, has recently joined the fry even in this more restricted area of operations, for the surprise of many observers.

In January 2017, the private–albeit partially funded by the state-owned China Aerospace and Science Industry Corporation (CASIC)—ExPace Technology successfully launched its first Kuaizhou-1 (KZ-1A[25]) commercial SLV carrying three small satellites. The company is now working on a larger and more competitive variant of the rocket, the KZ-11.

But private spaceflight in China started earlier than that, with LinkSpace, a Beijing-based company that was the first to receive private investments in 2014, even though so far has not gone further than prototype development of reusable rockets, of the kind of SpaceX. As Virgin Galactic, LinkSpace also aims at suborbital flights, and not just rockets (GB Times 2018).

Another company worth watching out for is the privately owned rocket-launcher LandSpace, founded in 2015 and based in Beijing. After successfully rising private capitals, the company announced in August 2018 the development of a solid-propellant carrier rocket, ZQ[26]-1, to send satellites into space. The first satellite scheduled for launch is Future, which is planned to remain on orbit for two years and collect data and remote sensing for China Central Television. The company is already developing another rocket, ZQ-2,[27] equipped with a liquid-oxygen/methane rocket engine, and expected to launch at some moment in the 2020s. "LandSpace's ascension is indicative of a broader trend in China today – the increased proliferation of private space companies. Launch companies alone include LinkSpace, OneSpace, iSpace, LandSpace, and ExPace (though ExPace is largely funded by CASIC and only nominally private). Among launchers, iSpace launched a suborbital rocket in April from Hainan, and OneSpace launched a suborbital rocket earlier this month. LinkSpace has also finished a vertical takeoff, vertical landing test in 2017. While launch companies have recently made the most headlines, other areas of space such as Earth observation, telemetry, tracking and command, and nanosatellites are also progressing in China" (*Space News*, 25 May 2018). It will not be long until these private Chinese launchers start to offer their services to the global market, making the sector far more diverse and competitive than ten years ago. This is just the beginning.

Case Study: Who's Who in the Global Space Sector

In addition to the general outlines of the commercial dimension of the space sector, in terms of its value chain and main segments, here's a few details on the most active companies[28]—some of which have been already mentioned—together with others, less famous and known by the general public.

A note about the choice of these twelve players: several companies not included in this roundup, which would otherwise qualify for dimension and important, have not been included because covered in other sections of this book. See the end of the case for more about the missing names.

Boeing

Country: USA

Sector: Aerospace and Defence

Size: a Fortune 500 company (#64 in 2017)

Revenues: US$93.39 billion

The largest company in aerospace and defence, Boeing participates in the satellite and launch components of the space value chain.

It provides launch services through the United Launch Alliance joint venture with Lockheed Martin to both commercial companies (Viasat) and the US government.

To notice: The company is developing CST-100 Starliner crew capsule which is a spacecraft design under construction by Boeing as their entry for NASA's Commercial Crew Development programme.

Corporate website: www.boeing.com

Lockheed Martin (LMT)

Country: USA

Sector: Aerospace and Defence

Size: a Fortune 500 company (#200 in 2017)

Revenues: US$51,048 million

LMT is one of the most important companies in the space industry, present in all segments of the value chain, with satellites (AEHF), launch (United Launch Alliance) and ground systems. It has a track record of a long working relationship with NASA and the US Defence.

To notice: Many recent investments to modernise its family of satellites, concluded in 2017 (Morgan Stanley 2018).

Corporate website: www.lockheedmartin.com

Northrop Grumman (NOC)

Country: USA

Sector: Aerospace and Defence

Size: a Fortune 500 company (#461 in 2017)

Revenues: US$25,803 million

NOC is another of the giants of the sector, present in all the segments. In addition to aerospace and defence operations, NOC also manufactures satellite payloads (e.g. the sensor system for SBIRS) and satellites themselves (like JWST).

To notice: It has recently entered the launching sector through the acquisition of Orbital ATK.

Corporate website: www.northropgrumman.com

SES Global

Country: Luxembourg

Sector: Satellites and Telecommunication

Size: Revenue €2035 million in 2017

The company is world's largest fixed satellite services operator (ahead of the American Viasat and Intersat), with 50 GEO and 16 Medium Earth orbit (MEO) satellites in service, providing TV and video communication to hundreds of million of people around the world.

To notice: The company has plans to increase the satellite fleet even further with 8 satellites in 2018/2019 and 7 in 2021.

Corporate website: www.ses.com

<u>**Inmarsat**</u>

Country: UK

Sector: Satellites and Telecommunication

Size: US$1400.2 million in 2017

The company is the leading mobile satellite services operator specialising in connectivity services for remote location, in-flight service and maritime routing.

To notice: Viasat has recently become a competitor in these segments, threatening Inmarsat's leading position.

Corporate website: www.inmarsat.com

<u>**Honeywell International (HON)**</u>

Country: USA

Sector: various

Size: a Fortune 500 company (#275 in 2017)

Revenues: US$40,534 million in 2017

Honeywell is a leader in avionics and life support systems, and it has historically equipped all for NASA's manned space mission. The company also has also expertise in manufacturing satellite hardware for military and commercial applications, providing guidance, propulsion, steering control systems to launch vehicles.

To notice: It has acquired in 2016 a Canadian satellite communication component manufacturer, Com Dev, to enter the small satellite market.

Corporate website: www.honeywell.com

<u>**United Technologies (UTX)**</u>

Country: USA

Sector: various

Size: a Fortune 500 company (#159 in 2017)

Revenues: US$59,837 in 2017

United Technologies' Aerospace System business is NASA's contractor in the segment of space suit (with its Extravehicular Mobility Unit—EMU) and life support system.

To notice: Among other things, it provides environmental control and life support equipment to the International Space Station.

Corporate website: www.utc.com

XL Group Insurance
Country: USA
Sector: various
Size: see below

XL Group is one of the top underwriters of space insurance, providing insurance coverage for launch and in-orbit events. Due to the growing of the segment of space tourism, the company will probably look to expand to the suborbital spaceflight market.

To notice: It has been acquired by the giant AXA in 2018.[29] The operation created the biggest property and casualty commercial lines insurer, with revenues of Euro 48 billion (£42.8 billion).

Corporate website: axaxl.com

CASC
Country: China
Sector: Aerospace
Size: a Fortune 500 company (#343 in 2017)
Revenues: US$34,254 million (2017)

The state-owned China Aerospace Science and Technology Corporation (CASC) is one of the main contractors for the Chinese space programme through its many subordinate entities. It designs and manufactures a range of spacecraft, launch vehicles and strategic and tactical missile systems.

Corporate website: english.spacechina.com/n16421/index.html

CASIC
Country: China
Sector: Aerospace
Size: a Fortune 500 company (#346 in 2017)
Revenues: US$34,073 million (2017)

The state-owned China Aerospace Science & Industry Corporation Limited (CASIC, Chinese; 中国航天科工集团有限公司) is a hi-tech company that is under the direct administration of the central government and another of the contractors for the Chinese space programme.

Corporate website: www.cccme.org.cn

ROSCOSMOS
Country: Russia
Sector: Aerospace
Size: 186.5 billion rubles (2015)

The Roscosmos[30] State Corporation for Space Activities, commonly known as Roscosmos, is a state corporation responsible for the spaceflight and cosmonauts' programme for the Russian Federation created in 1992.

To notice: It is currently developing a super-heavy rocket, to target both LEO and deep space missions, aiming at becoming operational in 2028.

Corporate website: en.roscosmos.ru

Mitsubishi Heavy Industries, Ltd.

Country: Japan

Sector: Space, Engineering and Electronics, Defence

Size: a Fortune 500 company (#311 in 2017)

Revenues: US$37,103 million (2017)

Mitsubishi Heavy Industries is one of the core companies of the Mitsubishi Group. After the technology transfer from JAXA (the Japanese Space Agency)[31] in the 2000s, the company took over the H-IIA—the Japanese workhorse rocket—launch service operations. Interestingly, albeit its impressive performance, aerospace is a small part of MHI, accounting from only 1% of its business.

To notice: The company is developing H3 rocket, a version without solid rocket boosters that is going to be cost-competitive with SpaceX's Dragon, and it is going to test it in between 2020 and 2021.

Corporate website: www.mhi.com

Sources

Most of the information listed here has been retrieved from the companies' corporate websites, listed at the end of each subsection, as of 16 December 2018. Additional sources have been indicated in the footnotes.

1. A separate section has been devoted to start-ups and recent initiatives in the launching and spaceflight sector in this Chapter 3 (one for all, SpaceX), while companies especially active in the segment of space tourism and suborbital flights (e.g. Virgin Galactic, Blue Origin) have been covered in Chapter 8, Sect. 8.3.
2. All the EU companies (such as Arianespace and its parent company, Airbus; BAE System; Leonardo; Alenia; Thales—just to mention the most representative) have been covered instead in Chapter 4, devoted to the European market.
3. The Indian Space Research Organisation (ISRO) acting both as a space agency and the leading country's manufacturers under its many entities has been included in Chapter 5, Sect. 5.3.

Notes

1. Such as SWOT analysis, PESTLE analysis, Porter's Five Forces, Porter's Diamond just to give a few examples. And yet, articles and books with analysis in this sense are still limited. As mentioned before, the space sector remains often confined to engineers, astrophysicists and rocket

scientists, while more expertise in the various aspects of the space sector as a whole is needed at this crucial moment.

2. This section is based on the conclusions of Gurtuna's seminal book on the space economy, to which it makes reference in more than one point. For more, see Fenema (1999), Nardini (2018).
3. This was exactly what happened for the satellite communication and that made Iridium a comparative failure when it finally launched in 1998.
4. Three main methods have been proposed: costing by analogy, parametric costing and bottom-up costing (Gurtuna 2013).
5. Space Shuttle costs are a good example; NASA put down a US$45 million estimates, but in practice the cost of each mission came down to US$1.44 billion (in 2010 dollars; NASA [2017]).
6. Risk management in space project is beyond the scope of this book. But an approach that mixes up portfolio diversification of missions and a built-in redundancy in the system would probably work better here as in other sectors.
7. A famous study that tried to assess space benefits is the one carried out by Midwest Research Institute (MRI) in 1971 and repeated in 1988. The study showed that US$207 billion invested by NASA translate in an overall return 7:1. In 1988 that ratio grew to 9:1. Another was carried out by Chase Econometric in 1975 using a more sophisticated approach, which yielded a ration of 14:1, even though attracted criticism. In 1978 also ESA engaged in some sort of evaluation of space benefits, obtaining somehow different results (Greenberg and Hertzfeld 1992).
8. There a few things worth noticing about this precious resource, whose periodic consultation is advisable to all space scholars. It is downloadable database which is periodically updated and can be sorted out by nation, launching sites and type of rockets. The statistics presented in this book have been retrieved from the official catalogue as for 22 November 2018.
9. UNOOSA is not the only organisation that monitors satellites. SIA does it as well. The US FAA maintains instead statistics of payloads, which can be retrieved at: FAA.org.
10. They are increasingly popular satellites, used not only by governments or big corporations but also by research centre and citizen-driven space initiatives. These rather peculiar characteristics make them one of the novelties of the latest years, which is contributing to change the face of space sector, and they are discussed in detail in Chapter 8.
11. This was achieved with 91 orbital launches in 2017, up from 85 launches in 2016 (Seradata SpaceTrak Database 2018); launches have been

discussed in detail in the next section. According to some provisional statistics for 2018 retrieved from UCS Satellite Database (2019), the number of satellites in operation as of 30 November 2018 was 1957, 849 of which operated by the US. About 1200 of them are in LEOs, making it the busiest Earth's orbit.

12. Statistics for the whole 2018 are not realised yet by SIA. The latest report is SIA's 2018 21st Annual State of the Satellite Industry Report (SSIR), released in June 2018 and available at: https://www.sia.org/2018_ssir/. SIA is a US-based trade association providing representation of the leading satellite operators, service providers, manufacturers, launch services providers and ground equipment suppliers. The reports themselves are prepared by Bryce Space and Technology.
13. The rest of the non-satellite space economy, amounting at US$79.3 billion in 2017, is composed by government space budget (about 24% of the grand total) and commercial human spaceflight.
14. As a matter of fact, it is only in 2016 that SIA-SSIR created a separate category for them, due to their growing importance.
15. A set of statistics of the launches can be found in the table and before starting examining them, a few methodological point needs to be made, especially what goes under the definition of "commercial flight". There is a quite precise definition from the FAA (2018), which gives some parameters for identifying a flight as such: "A commercial launch is a launch that is internationally competed (i.e., available in principle to international launch providers) or whose primary payload is commercial in nature. FAA-licensed launches carrying captive government (NASA and DOD) or industry payloads (ORBCOMM, Delta 3 demosat, Zenit 3SL demosat, and others) are counted here. Data are for orbital launches only". Moreover, the launch must have been privately financed without government support. Launch vehicles of commercial launches are manufactured and marketed by private companies, even when they carry government-funded payloads. In terms of regulations, commercial launches in the US territory have to obtain a licence from FAA.
16. Of these, Russia sent 13 (57%), Europe 6 (26%) and the USA only 4 (17%) of the total. This number has increased over time.
17. There are, as it often happens, a few discrepancies in terms of statistics depending on the source and how the calculations are made. Spaceflight 101 records, for example, 90 orbital flights, because 1 was not successful. When it comes to number of satellites, Seradata SpaceTrak gives a higher figure in its report, 466—which is the most spacecraft since

orbital spaceflight began in 1957. "The overall total was more than double the 223 spacecraft attempted to be launched in 2016 and over 150 more than the previous high of 302 in 2014" (source: Seradata SpaceTrack, 2017) https://www.seradata.com/final-score-for-2017-463-a-new-record-for-the-number-of-satellites-attempted-to-be-launched-in-a-single-year/. This is due to the fact that it also included lost or damaged satellites (37), or had their mission impaired by the launch vehicle or its underperformance (23 spacecraft did not achieve orbit at all). Classifications also vary when specific legal requirements are taken into account. For instance, FIAA reports instead 33 commercial space launches worldwide in 2017 (21 on 2016). The historical records so far is 35 carried out in the year 2000.

18. One was the Soyuz FG, which malfunctioned on its way to the ISS (but the crew was safely brought back to Earth). The other was the maiden launch of a Chinese rocket, Zhuque-1. In the table, partial statistics for 2019 have been provided.
19. If Dragon is a free-flying spacecraft, Falcon is "is a two-stage rocket designed and manufactured by SpaceX for the reliable and safe transport of satellites and the Dragon spacecraft into orbit. Falcon 9 is the first orbital class rocket capable of reflight" (SpaceX 2018). According to company's source, the 2018 variant—aka Falcon 9-Block 5—is going to be the final version. The design is going to remain the same in future years, while the development will focus on maximising reusability.
20. As a matter of fact, there is another spacecraft that could be used to carry astronauts, the relatively new Chinese Long March 2F (Space Daily 2018), adopted for manned flights since 2003. However, it has never been used, as China is not part of the ISS.
21. Just to give a rough idea of the figures involved, heavy rocket launches cost nowadays about US$90 million for SpaceX's Falcon Heavy, while NASA's Orion and its Space Launch System (SLS), still in its development stage, will probably about US$500 million to 1 billion for a single launch (its development costs are already about US$18 billion, according to some estimates; Space.com, 27 August 2018).
22. "Falcon Heavy is the most powerful operational rocket in the world by a factor of two, with the ability to lift into orbit nearly 64 metric tons (141,000 lbs)--a mass greater than a 737 jetliner loaded with passengers, crew, luggage and fuel. Falcon Heavy's first stage is composed of three Falcon 9 nine-engine cores whose 27 Merlin engines together generate more than 5 million pounds of thrust at liftoff, equal to approximately

eighteen 747 aircraft. Only the Saturn V moon rocket, last flown in 1973, delivered more payload to orbit," (SpaceX Press Release, 6 February 2018, online). The payload carried the Tesla Roadster playing David Bowie's tune Space Oddity, in what was criticised as a publicity stunt and a potentially dangerous endeavour.

23. It is under the NASA's Commercial Crew Development Programme. In one of the first missions, SpaceX will take the ISS astronauts from and back to Earth. Details of the programme, started in 2014, are available on NASA's page at https://www.nasa.gov/content/commercial-crew-program-the-essentials.
24. While Inspiration Mars (also called IM, an American non-profit that aimed to launch a crewed mission to Mars) looks no longer in activity, Mars remains on the radar of many private initiatives, as explored in Chapters 8 and 9. For more about IM, see https://www.nasa.gov/ames/ocs/2013-summer-series/taber-maccallum.
25. Albeit the private-developed KZ-1A is a light-lift quick response SLV for commercial uses, in some statistics is at times confused with KZ-1, its CASIC's military version, since they share a few aspects of design and operational characteristics.
26. Chinese rocket names are often lifted from the Chinese mythology. ZQ stands for Zhuque (i.e., Vermilion bird), which is one of the four spirits of ancient Chinese mythology, representing fire and controlling the South cardinal direction.
27. "Landspace officials said ZQ-2 will have the capacity to launch a 1500-kilogram payload into a 500-kilometer sun synchronous orbit, or 3600 kilograms into a 200-kilometer low Earth orbit. That would make ZQ-2 as much as 10 times more powerful than Electron, the dedicated smallsat launcher that U.S.—and New Zealand based Rocket Lab sent into orbit for the first time in January" (*Space News*, 25 May 2018).
28. This section has been compiled using a few open source reports freely available such as: Deloitte Report on the space sector 2017, Fortune 500 2018 and Morgan Stanley's (2018) investment report, together with the selected companies' corporate websites.
29. From the company website, press release, 2018, https://axaxl.com/insurance/news/axa-to-acquire-xl-group.
30. More details about Roscosmos partnerships to be found in Chapter 4, in Russia's country section.
31. Berger, E. (2018). Quietly, Japan Has Established Itself as a Power in the Aerospace Industry. *Ars Technica*. Retrieved 22 January 2019,

from https://arstechnica.com/science/2018/11/long-on-tradition-japan-grapples-with-a-rapidly-changing-rocket-industry/.

References

Business Insider. (2018, April 28). *SpaceX Cargo Launches May Soon Cost 50% More—But It's Still an Offer NASA Can't Refuse* [Online]. Available at https://www.businessinsider.com/spacex-nasa-launch-cost-increase-reasons-2020-2018-4. Retrieved on 28 March 2018.

European Commission. (2013). COM 2013/108 "Final EU Space Industrial Policy. Releasing the Potential for Economic Growth in the Space Sector".

FAA. (2018). *The Annual Compendium of Commercial Space Transportation.* Available at https://www.faa.gov/about/office_org/headquarters_offices/ast/media/2018_ast_compendium.pdf. Retrieved on 11 December 2018.

Fenema, H. P. (1999). The International Trade in Launch Services. In F. G. Von der Dunk & F. Tronchetti (Eds.), *Handbook on Space Law* (pp. 79–86). London: Edward Elgar.

GB Times. (2018, November 1). *Landspace Releases Preliminary Report on Chinese Private Rocket Launch Failure* [Online]. Available at https://gbtimes.com/landspace-releases-preliminary-report-on-chinese-private-rocket-launch-failure. Retrieved on 20 February 2019.

Greenberg, J., & Hertzfeld, H. (1992). *Space Economics* (Progress in Astronautics & Aeronautics). Available at https://arc.aiaa.org/doi/book/10.2514/4.866166. Retrieved on 28 April 2018.

Gurtuna, O. (2013). *Fundamentals of Space Business and Economics.* Westmount: Springer.

moonandback. (2013). *Dennis Tito and Team Outline Audacious Plan for Mars Trip | Moonandback* [Online]. Available at http://moonandback.com/2013/02/27/dennis-tito-and-team-outline-audacious-plan-for-mars-trip/. Retrieved on 28 February 2018.

Morgan Stanley. (2018). *Investment Implications of the Final Frontier.* Available at https://www.morganstanley.com/ideas/investing-in-space. Retrieved on 12 February 2019.

Nardini, F. T. (2018). *Space 2.0—The Commercial Space Revolution.* Adelaide: Fleet Space.

NASA. (2017). *NASA History Overview*. Available at https://www.nasa.gov/content/nasa-history-overview. Retrieved on 8 February 2018.

Neal, H. A., Smith, T. L., & McCormick, J. B. (2011). *Beyond Sputnik: U.S. Science Policy in the Twenty-First Century*. Ann Arbor: University of Michigan Press.

Norberg, C. (Ed.). (2013). *Human Spaceflight and Exploration*. Berlin: Springer.

Popular Mechanics. (2013). *The Rise, Fall, and Rise of U.S. Satellite Launchers* [Online]. Available at https://www.popularmechanics.com/space/rockets/a8811/the-rise-fall-and-rise-of-us-satellite-launchers-15222856/. Retrieved on 28 February 2018.

Seradata. (2018). *Seradata Space Trak Homepage* [Online]. Available at https://www.seradata.com. Retrieved on 28 April 2018.

SIA. (2017). *Satellite Industry Association's 2016 Report*. Available at https://www.sia.org/annual-state-of-the-satellite-industry-reports/2017-sia-state-of-satellite-industry-report/. Retrieved on 27 February 2019.

SIA. (2018). *Satellite Industry Association's 2017 Report*. Available at https://www.sia.org/2018_ssir/. Retrieved on 27 February 2019.

Space Daily. (2018, August 24). *Chinese Private Space Company to Launch First Carrier Rocket*. Available at http://www.spacedaily.com/reports/Chinese_private_space_company_to_launch_first_carrier_rocket_999.html. Retrieved on 10 November 2018.

SpaceX. (2018). *SpaceX Homepage*. Available at https://www.spacex.com/. Retrieved on 28 April 2018.

Space.com. (2018, August 27). *SpaceX's 1st 'Block 5' Rocket: A Tale of 2 Launches*. Available at https://www.space.com/41636-spacex-first-block-5-falcon-rocket-launch-future.html.

The Space Foundation (2019). *The Space Report 2018* [Online]. Available at https://www.spacefoundation.org/. Retrieved on 27 April 2019.

The Space Report. (2019). *The Space Report Online 2018* [Online]. Available at https://www.thespacereport.org/year/2018. Retrieved on 28 April 2018.

UCS Satellite Database. (2019). *Online Database*. Available at https://www.ucsusa.org/nuclear-weapons/space-weapons/satellite-database. Retrieved on 12 February 2019.

UKSA. (2018). *UK Space Agency*. Available at https://www.gov.uk/government/organisations/uk-space-agency. Retrieved on 11 February 2018.

Vedda, V. (2009). *Choice, Not Fate: Shaping a Sustainable Future in the Space Age*. Bloomington, IN: Xlibris Corporation.

4

Europe in Space and the Space in Europe: A Complex Relationship

4.1 A Strategy for Europe: The EU's Space Policy

The following official statement of the EU Commission in 2016 synthesises, better than several pages of discussion, what constitutes the heart of the matter in the European space sector: the presence of different and at time contrasting interests, especially on the security side, and the need of a more effective integration strategy to exploit synergies.

> Space services can strengthen the EU's and Member States' capacity to tackle growing security challenges and improve the monitoring and control of flows which have security implications. Most space technologies, infrastructure and services can serve both civilian and defence objectives. Although some space capabilities have to remain under exclusive national and/or military control, in a number of areas synergies between civilian and defence can reduce costs, increase resilience and improve efficiency. The EU needs to better exploit these synergies. (European Commission 2016b: 8)

This duality comes back to the early days of the European Community, since space was in Europe, as much as other sectors, the business of

S. Paladini, *The New Frontiers of Space*,
https://doi.org/10.1007/978-3-030-19941-8_4

individual countries and domestic policies. Nonetheless, it was as early as 1959 that the successful creation of CERN paved the way to a more comprehensive cooperation strategy on high-tech activities, and space was one of them.

Two organisations were set up in the following years, the European Space Research Organisation (ESRO) in March 1964 and the European Launcher Development Organisation (ELDO) in February 1964, to lead the two preeminent segments of the space industry at that time.

The academic literature identifies two factors that were fundamental in these regional institutions, namely (i) political will, which was decisive for the creation of the ELDO, and (ii) recognition by the scientific community of the stakes involved in space, which was instead the motivation behind the ESRO. The dual stream approach reflects the different states' mandates at the political level, in which military interests compounded the scientific focus on civilian applications and progress in research (Gaubert and Lebeau 2009).

These two organisations evolved, respectively, into the European Union Space Programme and the European Space Agency (ESA). ESA in particular, created in 1975, represents a unique case as a multinational space agency and, therefore, a working model for scientific and political cooperation at a regional level that has recently attracted a lot of attention.

ESA, its model and its activities are examined in more detail in the following section, but one thing needs to be clarified here: while working with the EU on a series of projects and receiving a substantial part of its functioning budget from the EU allocations, ESA is not an EU agency, but an independent multinational institution with its own budget, leadership body and mission.

As a matter of fact, the EU and ESA entered into a framework agreement as recently as in 2004, with the primary aim of developing a coherent and comprehensive European Space Policy (ESP).

As a result of this joint effort, a ESP was published in 2007 and represented the first attempt to establish one clear leadership of the sector and conciliate the two aspects of the European space strategy: the procurement approach of ESA, which is purely technical and commercial, with the political and military implications the EU keeps into consideration. This has not always been successful (Venet 2012; Venet and Baranes 2012), and it often has resulted into more or less open clashes, as, for example, in the case of Galileo (see case study at the end of the chapter).

Two executive bodies were tasked with the implementation of the ESP: (i) the European Council, which is responsible for the space framework of the EU, and (ii) the ESA Council of Ministers, which makes decisions on the space agency's programme and budget.

While this arrangement addressed the two "souls" of the EU space sectors in Europe, it also showed two potential problems.

One was (and it is just on the point to becoming worse) the imbalance in the membership of ESA and EU: not all the ESA members are also members of the EU, and not all the members of EU are members of ESA, which creates contrasting interests. The second is the autonomy of ESA itself, which has historically been a (much-overlooked) factor in steering the European space programmes into a direction or another.

With all these caveats, the ESP was adopted in a resolution at the fourth EU-ESA "Space Council"; later on, the Space Council Resolution in September 2008 defined "space and security" as one of four new priority areas. Finally, the Lisbon Treaty in 2009 set the legal basis for the EU's involvement in space matters in terms of one of the areas with shared competences with the EU's member states.

There are other areas of strategic policies that affect space matters, especially the ones that, for their dual-use characteristics, are part of defence and security agreements, e.g. the EU's Common Foreign and Security Policy (CFSP) and the Common Security and Defence Policy (CSDP). Both were adopted in the 1990s and were paramount for the deployment of civilian and military missions and the identity-building dimension. The CSDP missions have been historically confined to the EU's surrounding regions, defined in the EU's Global Strategy (EUGS) as building "State and Societal Resilience to our East and South" (Mogherini 2016), even though after 2008 those efforts have stalled for the lack of resources due to the global economic crisis.

But the importance of the space sector remained. The EU's High Representative Ashton, in her Final Report of October 2013 entitled "Preparing the December 2013 European Council on Security and Defence", reiterated those principles. In June 2016, the High Representative for Foreign Affairs and Security Policy and Vice-President of the European Commission, Federica Mogherini, presented the Global Strategy for the EU's Foreign and Security Policy. The 2016 Global Strategy "underlines the importance of the autonomy and security of European space-based services and the necessary investments in space systems for security, calling for improving the monitoring and control of flows which have security implications. This requires investing in Intelligence, Surveillance and Reconnaissance, including Remotely Piloted Aircraft Systems, satellite communications, and autonomous access to space and permanent earth observation" (EU-HR 2016, online).

Based on the Global Strategy, the EU released in 2016 two more documents directly dealing with space, security and defence:

A. The "Space Strategy for Europe"

European Commission, Communication from the Commission to the European Parliament, the Council, the European Economic and Social Committee and the Committee of the Regions on "Space Strategy for Europe", COM (2016) 705, Brussels, 26 October 2016.

B. The "European Defence Action Plan"

European Commission, Communication from the Commission to the European Parliament, the Council, the European Economic and Social Committee and the Committee of the Regions "European Defence Action Plan", COM (2016) 950 final, Brussels, 30 November 2016.

This second document looks particular important for what previews in terms of Space Surveillance and Tracking (SST) and Governmental Satellite Communication (GOVSATCOM), which are included in the precinct of the space budget allocation.

> The European Defence Action Plan (EDAP) [3] released in November 2016 contributes to ensuring that the European defence industrial base is able to meet Europe's current and future security needs and, in that respect, enhances the EU's strategic autonomy, strengthening its ability to act with partners. EDAP will promote civil/military synergies within EU policies, such as in the space sector, already highlighted in the Space Strategy. It also

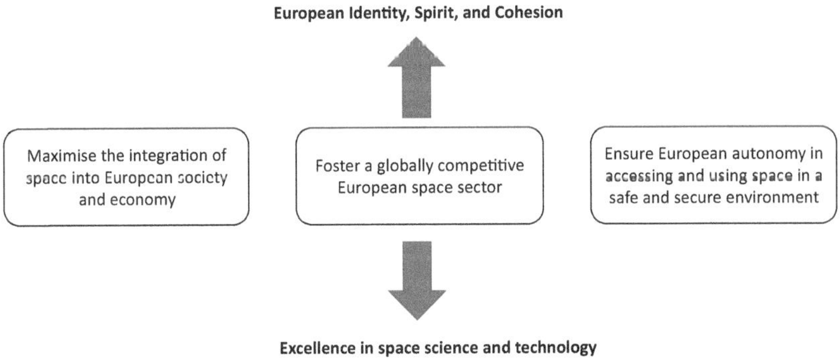

Fig. 4.1 European goals in space (*Source* Author's elaboration on EU-ESA data, 2016)

> confirmed the shift towards establishment of jointly developed security and defence space programmes, with renewed support for the SST programme and Commission involvement in the GOVSATCOM. (The EU 2016b: 10)

The strategic objectives are also reflected in the ESA/EU Joint Statement on "Shared Vision and Goals for the Future of Europe in Space" (ref. ESA/C/WG- M(2016)19) (Fig. 4.1).

The legal and institutional framework of the space sector[1] in the EU is now changing.

After the UK's referendum over Brexit in June 2016 and the decision of invoking article 50 in March 2017, the EU reacted by launching European Defence Fund (EDF) as part of the Commission's European Defence Action Plan. On 6 June 2018, the European Commission presented its new space programme,[2] which builds on the "Space Strategy for Europe" of October 2016 and on the "Industrial Policy Strategy" presented by President Juncker in his 2017 State of the Union address.

It also increased the EU space budget, a trend that has been constant over a few years.

And if the budget 2014–2020—part of Horizon 2020—committed over EUR 12 billion in space activities, the funding approved in June 2018 allocates 16 billion Euro to the sector, divided according to the EU's ongoing three flagship programmes and other strategic priorities:

A. Copernicus, the leading provider to Europe in the EO (Earth observation) segment. This system is used, among others, for atmosphere and marine monitoring, disaster management, climate changes and agricultural management objectives.
B. EGNOS, the European Geostationary Navigation Overlay Service, which augments the GPS signals over Europe and provides "safety of life" navigation services to aviation, maritime and land-based users over most of Europe.
C. Galileo,[3] the new European Global Navigation Satellite System (GNSS), due to be fully operational in the mid-2020s.
   Galileo and EGNOS are going to receive €9.7 billion in fund allocations, Copernicus €5.8 billion.
D. The new EU security concept, which includes the performance and autonomy of Space and Situational Awareness (SSA) and the new GOVSATCOM, which grants secure satellite communications to national authorities and monitors space hazards. This part of the programme is going to receive €500 million, to be devoted to grants for collaborative research into innovative defence technologies and products.
E. Finally, in a provision that has left many observers puzzled, the EU new space strategy announced the creation of the EU Agency for the Space Programme. In practice, the provision will increase the budget and the field of activities of one of the many EU agencies that deal with space at one title or another, namely the Prague-based European GNSS Agency (GNSS Agency or GSA).

The measure goes into an opposite direction of what stated by the Joint EU/ESA Statement signed by the two parts in December 2016. In this joint statement, ESA and the EU agreed on a number of common goals, highlighting the needs of stronger cooperation in the future and defining ESA as "THE" space agency of Europe. Therefore, the emphasis on GNSS came out as unexpected. This relatively small EU body set in the Czech Republic is more a bureaucratic than a technical body, managing public interests in the European GNSS programmes, European Geostationary Navigation Overlay Systems (EGNOS) and Galileo, the European global satellite-based navigation system.

In the outcry that has ensued, the EU Commission denied it was set on creating a rival to the ESA, reacting strongly to claims made by Jan Wörner, ESA's director general, that the EU 2021–2027 plan for space research amount to a blueprint for a duplicate body. It will "*take decades and cost billions*", Wörner said (*The Guardian*, 6 June 2018). In her speech, the High Commissioner for industry Elżbieta Bieńkowska has declared ESA will remain "the main partner in programme implementation, given its unmatched expertise. There is definitely no transfer of competence from the European Space Agency to the future Agency for the Space Programme. [...] All the institutions need [each other]. We are just strengthening the EU agency for the programmes to adapt itself to the new space activities and the security accreditation" (BBC News, 6 June 2018).

Nonetheless, it was inevitable that the budget allocation was to be regarded as a way to strengthen the EU grasp on the space policy and, on the other hand, to advocate competencies formerly residing with ESA in an attempt to diminish ESA autonomy (Science Business 2018).

Whether this can be achieved, or even attempted, it is another story.

4.2 The Space Sector in Europe

The substantial growth of the European space sector is reflected in the EU documents released over the years. The evolution is evident[4] in terms of volume of sales, investments, but also in quality and market integration.

However, a closer look at the sector shows that it is still difficult to speak of "a" European space industry. The institutional market is fragmented due to the existence of many public bodies and domestic companies active in each of the member states. Therefore, any analysis has to take into account both the general outlook and the reality of the single countries. Given its importance, ESA has been treated into a separate section, and so has the UK, which seems on its way to leave the EU but that remains one of the fundamental countries in the European space sector nonetheless.

According to the latest data, the EU's space industry in 2017[5] was estimated at €53–62 billion in added value, i.e. the second largest in the world after the USA and employing over 231,000 personnel. Europe manufactures one-third of all the world's satellites and maintains a dynamic downstream sector with a large number of SMEs in satellite and consumer services.

According to Eurospace, a consultancy that monitors the European space sector on a yearly basis, in 2017 sales were worth €8.76 billion (+6.2% compared to 2016). The demand from domestic markets (grown by €703 million) proved to be stronger than foreign customers, which showed a decrease in exports (€ −188 million; Eurospace 2018).

As in previous years, the institutional programmes promoted by the EU, ESA and the member states accounted for more than half of the European business (59% in 2017, amounting to €5.1 billion). The remaining 41% is represented by the commercial segment (€3.5 billion), with commercial satellites accounting for €2 billion, mainly telecommunication systems (€1.7 billion) and EO (€0.2 billion).

For what concerns the launching industry, Europe built two of the most reliable space vehicles still in service: one is the Russian Soyuz, the only remaining spacecraft to take the astronauts into orbit; the other is Ariane 5 (part, together with Vega, of the Arianespace fleet[6]).

The launch segment amounted to €989 million in 2017 (up from €873 million in 2016). Incorporated in 1980 in France, Arianespace[7] is a private company with the mandate to market Ariane launch vehicles and offer launch services on a commercial basis (Iserland 1990; Baudin 1999). Arianespace is one of Europe's success stories since its onset almost forty years ago, when it managed to capture about half of the world's launch market share within its first ten years of existence (Horsley 2005). Its heavy rocket Ariane 5, first flown in 1996 from its launch pad in French Guyana, is the ESA's workhorse and the main carrier of all-European missions.

Ariane 5[8] has flown to date more than 100 mission and put more than 200 satellites into orbit, including Galileo navigation system, even though it attracted at times criticism and even lawsuits by its competitors; for example, on 25 May 1984 the US private launch service provider Transpace Carriers Inc. (TCI) filed a complaint against Arianespace, allegedly due to subsidies Ariane had received by ESA and some EU member states (Fenema 1999; Krige 1999).

Lawsuits apart, the rocket's reliability is one of the highest of the sector, with 82nd consecutive successful launches by December 2017 (ArianeGroup 2017). Ariane 5 is trusted also by non-European contractors with

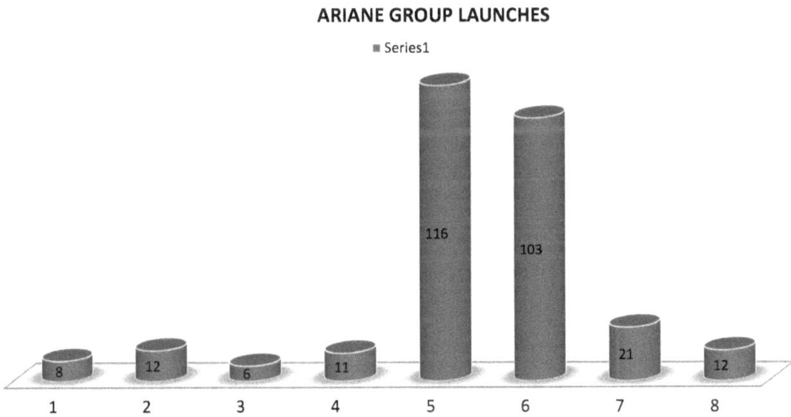

Fig. 4.2 Arianespace launches by typology of rockets (*Source* Author elaboration on ArianeGroup data, 2019)

projects as important as, for example, the new NASA's James Webb Space Telescope, which will be launched by an Ariane 5 rocket on 2021 (NASA Press Release 2018). Arianespace and Ariane 5 launched on 5 December 2018 the tenth mission of the year, putting on orbit two satellites, one for India (ISRO) and the other for South Korea (Fig. 4.2).

An example of the plurality of opinions in the European space sector—and inside ESA—is the level of debate about the future of Arianespace and, more specifically, the development of the next generation of European launchers. Germany rooted for the improvement of the existing models, while France supported the project of the new Ariane 6, which eventually gathered widespread support. The cost of developing the new launcher was estimated at four billion euro, with an expected completion timeline of about eight years (*SpaceNews*, 8 September 2014). Ariane 6 is going to eventually replace Ariane 5, and its estimation is to be 40% cheaper than its predecessor (Space Daily, 26 September 2018). This is due to the pressure put by SpaceX in terms of competition over cost and pace of space missions.

Another long discussion, which lasted until June 2018, was about the successor of the light launcher Vega, called Vega C. Since both Ariane 6 and Vega C[9] are more flexible than the two workhorses that are going to replace, there was more than a possibility of clashes between Italy and France regarding which of the two rockets should be given priority and government funding (*Les Echos*, 16 June 2018). Eventually, ESA agreed to support and promote both Ariane 6 and Vega C rockets with their prospective customers, even though the modalities and details are still unknown.[10] The transition to Ariane 6 is expected to be completed by 2020, when the heavy rocket is scheduled to have its maiden flight.

It is not by chance that all the discussions about the future launchers have been between Germany, France and Italy. The three states account for most of the investments in this segment. France, in particular, was one of the first countries to start a space programme, with scientific research, the domestic development of rockets in the 1960s, and the creation of its own space agency (CNES) in 1961. In 1973, France put forward the proposal of a ESA, to which it became the first contributor. The French government commits to space the largest budget in Europe and the second in the world after the USA (€2.33 billion[11]; CNES 2017).

This is partly explained by the presence of a robust domestic industry (ArianeGroup being only one among other space operators; among other well-known companies, there is Ariane's parent companies, Airbus, and then Safran, Astrid, Thales, Aerospatiale and Dassault) but also by the spaceport from which all the ESA rockets launch. Kourou, in French Guyana, is an ideal launching site due to its nearness to the equator. The location allows profiting from the Earth's spinning, saving propellant and also reaching polar orbits.

In addition to that, France is involved in international partnerships beyond ESA, especially in the field of space exploration. Among recent endeavours, the cooperation with NASA on Mars InSight (successfully deployed on Mars in November 2018) and the Solar Orbiter launched in Summer 2018 towards the Sun. Other partnerships have seen French companies participate in Chinese, Russian, Indian and Japanese projects.

At times strategic partner, at times competitor, Italy works closely with France on a series of initiatives, first of all the development of the new ArianeGroup's launchers mentioned above. Another important example is Space Alliance, a partnership started in 2005 and that puts together two of the countries' major groups, the Italian Leonardo (i.e. the former conglomerate of defence-aerospace Finmeccanica) and the French Thales.

From that strategic partnership, two joint ventures were formed: Telespazio (67% owned by Leonardo, 33% Thales) and Thales Alenia Space (67% Thales, 33% Leonardo).

Telespazio is one of the world leaders in satellite services, from launch support services to telecommunications to navigation and EO. The company is one of the major providers of Galileo, EGNOS, Copernicus, GMES and COSMO-SkyMed and recorded €577 millions of sales in 2016. Thales Alenia Space is another world leader not only in satellite services but also in science and research, with a 2016 global revenue of €2.6 billion.

The Italian space sector, albeit successful and well developed,[12] has recently undergone a substantial transformation,[13] both in its sector holding (Leonardo-Finmeccanica) and in the national space agency, ASI (with the new law on space "DDL Spazio" in January 2018; ASI 2018). However, with the entrance in the stock exchange of AVIO Spa and the start of the development of Vega C, the outlook for the domestic space sector looks good.

The German landscape is somehow more complicated to analyse. While the country has been the largest contributor to the ESA's budget since the onset and its two aerospace companies, Airbus and OHB, were always involved at some title in all-European space projects, the country lacks that dynamism in the private sector which is characteristic of other European countries, the UK first. "In Germany, we don't have enough young people who want to run their own business. We have an overall lack of high-tech companies and so it is in the space sector as well", a DLR's senior executive declared (*SpaceNews*, 2017, online). This is why DLR, the German national space agency, has recently started a space programme of incubation centres and project competitions of the kind NASA has been doing for decades (see Chapter 5).

One of the reasons for this difference in approach is due to the lack of venture capital mentality[14] and culture especially in the space sector, which seems to be an overall feature in continental Europe.

4.3 The European Space Agency (ESA)

It's impossible to overestimate ESA's importance.

Founded in 1975, ESA[15] is the leading operator in the European space programmes, as a recipient of funding provided by its 22 member states, EUMETSAT and the EU. To have a rough idea, ESA represented in 2017 the major space contractor, with 40% of total industry sales and 67% of European institutional customers. With more than 20 satellites currently in orbit, over 30 planned in the next 10–15 years, and space vehicles in development—notably the future European-built launchers Ariane 6 and Vega C—it constitutes the largest institutional provider of space sector services in Europe.

But the importance of the space agency goes beyond these figures. It has already been pointed out that ESA is an international organisation independent from the EU,[16] albeit it receives from the EU member states 4/5 of its funding (Fig. 4.3).

ESA's annual budget falls in two categories, "mandatory" and "optional". Space science programmes and all the related activities belong to the first, to which all member states contribute based on their internal GDP. The others—such as EO, telecommunications, satellite navigation, the International Space Station and microgravity research—are instead "optional", and, since they are not of interest to everybody, each member state is free to contribute or not (ESA 2018b).

The procedure of budget approval in ESA is different from, say, NASA, which is an annual exercise; ESA mechanism works instead on a multi-period system, with financial decisions on operative matters every two to three years and long-term funding cycles of 10 years (ESA 2015). Member states' delegates take long-term strategies to the ESA Council of Ministers. As it has been noted, the system proved quite effective in the occasion of the global economic crisis of 2008. "As the 2007–2016 long-term plan protected funds allocated to space agency projects, and despite larger economic woes, this formula allowed an increase in the agency budget" (Giovanni Bignami, Chairman of ESA Space Science Advisory Committee, as quoted by ESA 2005, online). As a result, ESA Ministerial Council meetings in 2012, 2014 and 2016 have increased ESA operating budget year by year in the latest period (Fig. 4.4).

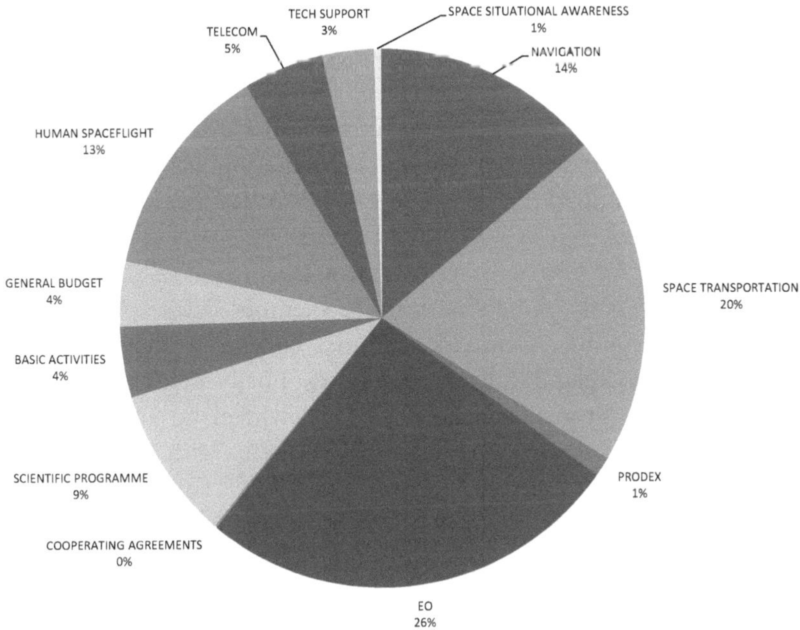

Fig. 4.3 ESA budget 2018 by domain (*Source* Author's elaboration on ESA data)

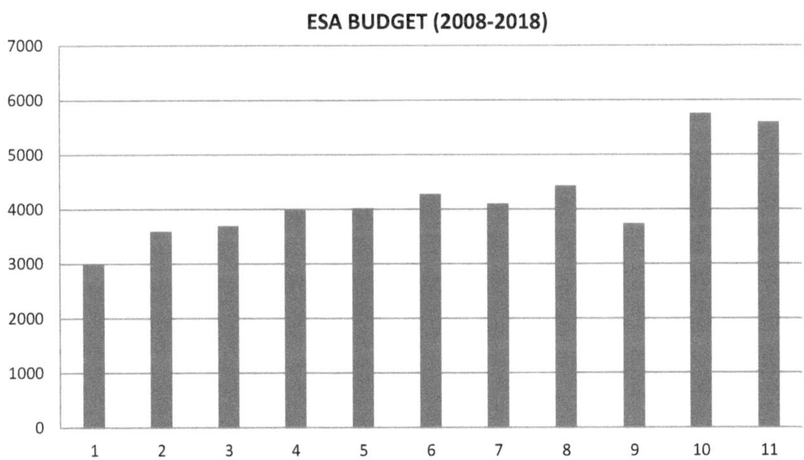

Fig. 4.4 ESA budget (2008–2018) (*Source* Author's elaboration on ESA data, 2008–2018)

Among them, France and Germany are historically the largest contributors (respectively 24.2% and 23.1% of 2018 total budget allocation of €5.6 billion), followed by Italy (11.8%), the UK (8.4%), Spain (5.2%) and Belgium (5.1%). It is important to remember that ESA's geographic *juste retour* policy make sure that the contributing states get back the ratio of their funding to the agency in value of contracts awarded to their national industries. This is one of the reasons why many consider ESA a "safe investment for new agencies" (Table 4.1).

ESA's reach is truly global. With its 22 member states—Austria, Belgium, Czech Republic, Denmark, Estonia, Finland, France, Germany, Greece, Hungary, Ireland, Italy, Luxembourg, The Netherlands, Norway, Poland, Portugal, Romania, Spain, Sweden, Switzerland and the UK—and a series of international cooperation agreements with the main space agencies in the world (NASA, Roscosmos and JAXA among them), ESA carries out activities such as Earth observation, human spaceflight, space transportation, navigation, science, engineering and technology beyond the capacity of any individual member state (ESA 2018a).

It has been observed that,[17] compared to other space agencies, ESA seems more focused on the economic aspects of space activities than on exploration of outer space or even more visionary aims of space colonisation. While this can appear somehow unfair in views of some recent ESA's success in terms of inspirational missions (e.g. Rosetta), there is some truth in this claim when looking at the kind of missions ESA supports.

Among the most important, there are the EU's flagship missions, Galileo, Copernicus and EGNOS, which are co-funded[18] by the EU.

Copernicus is one. The EU started the Copernicus mission as part of Horizon 2020 in partnership with ESA, with the aim of developing a global Earth observation capability for environment and security. Copernicus provides free, open data access to thousands of users in the public sector, the research and scientific community, and businesses. As part of this programme, ESA directly manages seven satellites (the seventh was launched in April 2018), together with elaborating data coming from 30 partner space agencies.

Copernicus is already the third largest provider of EO data in the world, but its capabilities will grow even further when more "families of satellites", the so-called sentinels[19] will be added to the constellation by 2021. They

Table 4.1 Current and future ESA missions (2018)

Name	Purpose	Description
Current missions operated by ESA		
Aeolus	Earth observation	(1) ESA's wind mission
BepiColombo	Planetary science	(2) Mercury explorer
Cluster II	Science	(4) Earth's magnetic environment
CryoSat	Earth observation	(1) ESA's ice mission
ExoMars/TGO	Planetary science	(1) Orbiting Mars—delivered the Schiaparelli entry, descent and landing demonstrator
Gaia	Science	(1) Galactic surveyor
Integral	Science	(1) Gamma-ray observatory
Mars Express	Planetary science	(1) Europe at Mars—delivered the Beagle lander
Sentinel 1	Earth observation	(2) Europe's Copernicus programme—Sentinel-1A, 1B (radar)
Sentinel 2	Earth observation	(2) Europe's Copernicus programme—Sentinel-2A, 2B (optical)
Sentinel 3B	Earth observation	(1) Europe's Copernicus programme—3B now operated by ESA until end of commissioning in Autumn 2018, then handover to Eumetsat
Sentinel 5P	Earth observation	(1) Europe's Copernicus programme—Sentinel-5 Precursor (atmosphere)
Swarm	Earth observation	(3) Geomagnetic surveyor
XMM	Science	(1) X-ray observatory
Proba 1, 2, 3	Earth observation	ESA's first microsatellite with 'smart' processing and self-control
SMOS	Earth observation	ESA's Soil Moisture and Ocean Salinity (SMOS)

(continued)

Table 4.1 (continued)

Name	Purpose	Description
Future missions operated by ESA		
Biomass	Earth observation	Measuring forest biomass
EarthCARE	Earth observation	ESA's cloud, aerosol and radiation mission
Euclid	Science	Mapping the dark universe
ExoMars RSP	Planetary science	ExoMars rover and surface science platform
Flex	Earth observation	Fluorescence Explorer to map vegetation fluorescence and quantify photosynthetic activity
Juice	Planetary science	Exploring the Jupiter system
Plato	Science	Discover extrasolar planetary systems and terrestrial planets
Sentinels/Copernicus	Earth observation	Europe's Copernicus programme—Sentinel-6A and -6B
Solar Orbiter	Science	Solar and heliospheric physics
Cooperative Missions operated by others		
ISS—International Space Station	Orbiting science lab	NASA (USA)/Roscosmos (Russia)
SOHO	Solar observations	NASA (USA)
Hubble Space Telescope	Orbiting observatory	NASA (USA)

Note The number in parentheses refers to the number of satellites in the mission
Source Author's elaboration on ESA public data, 2018

have enhanced capabilities such as radar imaging, high-resolution super-spectral imaging or altimeter capabilities, which will allow the EU to data monitoring of the territory, marine environment, atmosphere and help organising emergency response, security and climate change. Most importantly, data are going to be accessed "by EU agencies as well as most member states' national government agencies. The cost of the programme, about €4.3 billion until 2020, is divided between the EU and ESA on a two-thirds to one-third basis" (*FT*,[20] 2017, online).

The cooperation between ESA and the member states and ESA's relationship with the EU's various institutions is particularly complex (Moslinger 2017; Suzuki 2003), and it has been briefly examined in the previous section.

There is another aspect, which is not often considered when discussing ESA's activities, and it is the one related to security, both traditional and non-traditional, in which ESA has been remarkably active.

ESA's commitment descends directly from the agency's Convention Purpose, which is "to provide for and promote, for exclusively peaceful purposes, cooperation among European states in space research and technology and their space applications" (Art. 2).[21]

Starting with its Long-Term Plan (LTP) 2007–2016, ESA has included a various range of security activities, in particular in EO, navigation and telecommunications, setting up a comprehensive regulatory framework to deal with the threats identified in its Security Agreement and Security Regulations.[22]

As Giannopapa states, "the increasing need for security in Europe and for safety and security of Europe's space activities has led to a growing number of activities in ESA in various domains. It has also driven new and strengthened partnerships with security stakeholders in Europe. At the European level, ESA is collaborating closely with the main European institutions dealing with space security, namely the European Commission (EC), the European External Action Service (EEAS), the European Defence Agency (EDA), the European Union Satellite Centre (SatCen) and the European GNSS Agency (GSA)" (Giannopapa et al. 2018: 2) (Fig. 4.5).

There are a series of texts and agreements developing ESA's positions on security issues, but one of the most important is "Elements of ESA's Policy

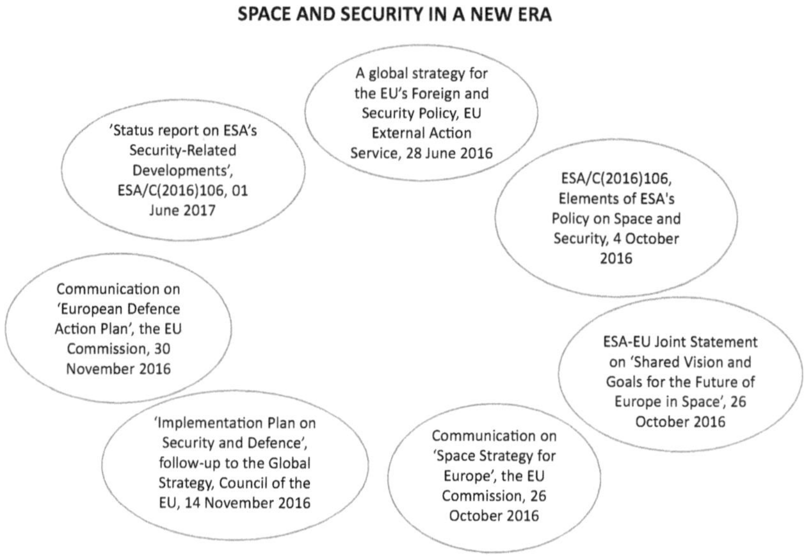

Fig. 4.5 Space and security in a new era (*Source* Author's elaboration on EU-ESA data, 2018)

on Space and Security", ESA/C(2016)106, adopted in ESA Council of October 2016. Among the guidelines agreed, there is ESA's aim at carrying out actions to ensure security on a comprehensive approach.[23] On 25 October 2018, ESA established four pillars for its activities, starting the negotiations for a new Financial Framework Partnership Agreement with the EU for the 2021–2027 period, where security protocols and procedures are going to play a fundamental role. The four pillars are safety and security; space exploration, applications (Earth observation, navigation or telecommunication) and enabling and support (launch capacities, technologies).

One area in which ESA has been particularly active is SSA—"Space Situation Awareness[24]"—which covers non-traditional threats, and it is split into three main segments: Space Weather (SWE), which entails monitoring the Sun and the solar wind to prevent damaging space-borne and ground-based infrastructure or endangering human life; near-Earth objects (NEO), detecting natural objects, such as asteroids and comets, which can potentially impact Earth; SST, tasked with tracking active and inactive satellites and space debris (collectively these items are referred to as Resident Space Objects—RSOs). SSA SST is carried out by radar and optical surveillance[25] across member states at a few different locations. More specifically, for what concerns debris, in terms of both mapping and mitigation initiatives, ESA carries out one of the most advanced programmes in the world, illustrated in Chapter 7.

In addition to the existing missions, there are new initiatives on the table, such as an agreement between the Italian Space Agency and ESA that will provide Europe with a detection system (the FlyEye telescope, which is expected to become operational in 2019). A joint mission ESA and NASA (HERA) follows NASA's Double Asteroid Redirection Test in its study of the 800-metre Didymos asteroid.

4.4 The UK: A Space Power Between the EU and ESA

No exam of the leading European space active countries is complete without the UK:

> The UK space sector punches above its weight globally and continues to grow much faster than the UK economy. Space is a high growth, high productivity sector where targeted government investment generates further industrial investment and creates highly skilled jobs. It is a key enabler that is critical for our national economy. (UKSA 2017, online)

The above claim is not preposterous: the UK has a long and distinguished tradition as a space power. After a space programme started as early as 1952, the country became the third nation after the USSR and the USA to launch a satellite into orbit back in the 1960s (Ariel 1, in 1962), and since then, it has been a leading player especially in sectors as EO, small satellite manufacturing and robotics.

The country has also an extensive programme of international cooperation, among others, with NASA. The UK is one of ESA's[26] largest contributors, participating in all its flagship missions, both Earth-bound

(Copernicus, Galileo) and space exploration (BepiColombo, Euclid, ExoMars Rover).

The UK space sector has been recently restructured with the creation of the UK Space Agency (UKSA)[27] in 2011 (Mosteshar 2013). The agency, which is part of the Department for Business, Energy and Industrial Strategy, is in charge of the national civil space activities. In a recent study, the UK government explains in detail the priorities of the national space programme. "Particular aims include coordination of space activity, supporting industry, encouraging academic research, increasing understanding of space science, and inspiring the next generation of UK scientists and engineers. The Agency's work with government, industry, business and academia has created major successes for the UK industry and economy - experiencing a decade-long growth trend in the space sector. Continued focus is given towards cooperation and participation in the European Space program" (the UK Government 2018, online).

Another action was the establishment of "poles of excellence", such as a space cluster in Hartwell, the Satellite Applications Catapult and the ECSAT (an ESA body).

Another recent development has been the approval in March 2018 of the Space Industry Act, an important milestone in commercial spaceflight activities, which include vertically launched rockets, sub-orbital spaceplanes, satellite operations and the construction of spaceports in the UK, together with representing one of the essential legal components for the development of a space tourism sector.

Also thanks to these proactive policy measures, the UK space sector has kept growing at an average of 8% since 2000, which is five times greater than the country's economy in the same period. According to the latest data released by the UKSA (2018), the KPIs are impressive: in 2017, the UK space industry was worth £15 billion a year in terms of turnover, employing more than 38,000 people across the country, while its exports accounted for £5.4 billion. And if at present the UK's share of the global space sector is about 6.5%, the aim is to reach 10% by 2030.

A more accurate breakdown of the sector is provided by the latest UKSA Report (2016), which gives the overall value of the sector at £13.7 billion for 2014/2015, with internal figures as follows:

- the productivity of the sector was about 2.7 times the UK average in 2014;
- the upstream segment accounted for £1.7 billion;
- £12.0 billion went to the downstream segment, with its space applications leading the Table (74% of the total), followed by space operation (15%), space manufacturing (8%). Ancillary services amounted at about 3% of the total. Broadcasting remained the leading subsegment (56%) but in a measure more limited compared to previous years, followed by communication (20%) and navigation (12%);
- exports were sustained (£5 billion in 2014/2015), and the sector performance as a whole was about 30% higher than the rest of the country. The first destination of the UK export was Europe, with 49% of the total;
- in terms of geographic location on the national territory, not all the regions contributed in the same way to the space sector. London was one of the major poles of attraction (about 68% of the sector's value chain), but other areas were the South and East of England, the last one especially in terms of space manufacturing. The three areas alone accounted for about 94% of the UK's total space income;
- interesting insights were given by the breakdown of the workforce profile. Only four areas were over the UK average (0.12% of the total workforce), London and South-East (0.21%, respectively), East (0.17%) and Scotland (0.27%). The odd case of Scotland, which registered a share of employment (18%) larger than income (0.5%), was due to the fact that a few UK organisations headquartered elsewhere—e.g. in London—maintain their operation bases elsewhere, for instance in Scotland (as the DTH broadcasting call centre);
- the government support was more limited compared to other countries (the world average has been estimated at 24%). The UKSA only contributed 3.1% and the government as a whole with its various organisations about 10.8%. The industry was commercially led and highly concentrated, with less than 20 companies accounting for 89% of the global income.

This industry profile is still valid as of today, especially for what concerns the UK companies. Some of them are world-famous aerospace giants, e.g.

Airbus[28] and BAE Systems,[29] others,[30] such as Surrey Satellite Technology Ltd. (SSTL), manufacture cutting-edge technology such as the payloads for the Galileo navigation service.

Among the UK leading companies in the space sector, there is Lockheed Martin UK, which, together with being the fifth largest supplier to the Ministry of Defence, also spends about £1 billion every year supporting about 1000 companies in the supply chain; it has recently entered into partnership with the UKSA to build a commercial spaceport.

The upstream segment represents one of the most interesting developments for the UK space industry as a whole, since a commercial spaceport can support launch services for small satellites, an area where the UK is particularly competitive, and the fledging space tourism sector. The regulatory framework was created, as mentioned above, in March 2018 with the Space Industry Act, and a case has been made on its feasibility.

In a study commissioned by the UKSA to Frost and Sullivan (2018), it was estimated that the small satellite market value in the UK would amount at US$5.5 billion in the period 2021–2030 (the world estimate is given at US$115.9 billion), with a number of satellites equal to 17,374 (2018–2030). The reason for this staggering growth is due to the fact that now many commercial operators work with "constellations" of small satellites instead of relying on bigger ones as in the past. About 70% of them will be in the category of <250 kg,[31] with 7217 of them launched into LEO and therefore with a limited lifespan (inferior to five years).[32] The potential market is huge, considering that the existing launching capability for them can only meet about 35% of the demand. Moreover, since small and nanosatellites are always launched as a secondary payload on top of other missions, they have always to compromise with bigger ones for required orbit and waiting time.

The development of spaceports in the UK will help meet this potential demand, and a few locations have already been identified by the UKSA. Moreover, £50 million have been earmarked by the Government's Industrial Strategy to help develop a domestic launch programme. Among the locations identified, there are seven sites from Cornwall to Scotland, including Glasgow's Prestwick. In addition to spaceports, which are going to allow a low-cost access to space, other areas of the UK targeted growth by 2030 (about £75 billion; UKSA 2018) have been identified in EO, connectivity services, in-space robotics.

4.5 Russia: From Sputnik to Roscosmos

The reason to include Russia in a chapter devoted to Europe is not just due to geography. Since the collapse of the Soviet Union, Russian public and private space operators have been increasingly involved with their European counterparts, even though the Russian space programme[33] itself has gone through a severe crisis of governance and dwindling resources in the first decades of the new century. And while the worst seems over, the programme is now only a fraction (about US$4 billion in 2015) of what it used to be at the time of the Soviets, when the Russian dominated the space together with the USA.

It is a fact that no other country, with the only exception of the USA, has done as much as Russia in terms of space programmes, pursuing all the fields from research to industry to space exploration, with some notable achievements. After all, the Russian have carried out more space launches than any other country in the world since the development of the first ICBM in the 1950s, surpassing, at some moment, the rest of world in terms of volume.

Nothing shows the prowess of Russian science and technology better than the Soyuz. This robust, medium-lift, four-stage launcher is probably the rocket with the highest reliability in use today, going through a long series of evolutions since the launch of Sputnik. In sixty years of service,[34] the Soyuz has been used without interruption, making more than 1800[35] between manned and unmanned missions. The Soyuz is not the only example of the longevity of the Russian technology: for example, the USA still adopts a Russian rocket engine, the RD-180, on one of its main space launch systems, the Atlas V (Fig. 4.6).

By looking at the statistics, the watershed represented by the fall of the Soviet Union is evident in the space programme as much as in the other sectors of the Soviet economy.

Upon the breaking up of the USSR, there was a temporary stop of investments in space activities, the dissolution of the Soviet-led space agency and

Fig. 4.6 Soyuz

the institution of national space agencies in the new-born CSI republics. Kazakhstan came first in 1991, followed by Ukraine in 1992 and Russia in 1992 (the Federal Space Agency). This constituted a novelty compared to the original Soviet design, which had no central executive agency for the space programme, but many different bureaux. This also explains the initial challenge for the agency in terms of governance.

The 1900s were difficult, with severe issues of funding, and Russia struggled to manage its space station (MIR), to contribute to the ISS and to carry out any other initiative. In order to deal with the domestic shortage of capitals, the agency tried to find alternative sources of funding, by providing its launching services to the commercial satellite markets and exploring space tourism, while carrying out only limited space exploration.

The improved conditions of the Russian economy in the 2000s made more money available to the space programme (the ten-year budget 2006–2015 committed about US$11 billion to the space agency, plus more money to the single missions). In those years, the space agency pursued a few priorities, such as the GLONASS navigation system, the development of the new Angara rocket and the participation to the ISS, which, alone, consumes about half of the entire Russian space programme budget.

After some especially bad years in 2010–2011—with the failure of a Prokton rocket that sent three satellite GLONASS into the sea, the partial failure of the Rockot launch of a military satellite and the disaster of the space exploration mission Phobos—the sector was reorganised with the incorporation of the United Rocket and Space Corporation as a public company. The Russian government also decided a re-nationalisation of the space sector to improve the reliability of operations and the overall efficiency.

The latest act was the consolidation in 2015 of two separate organisations, the Russian Federal Space Agency and United Rocket and Space Corporation in one single entity called Roscosmos State Corporation for Space Activities (Roscosmos), which is the space agency in operation as of 2018, together with being one of the main Russian rocket and satellite manufacturers.[36]

Russia operates the third largest fleet of spacecraft including communications, meteorological and reconnaissance satellites (Aerospace Technology 2015), 160–180 of which for scientific and military purposes (Tarasenko 1996).

In terms of the launching market, Russia's traditional dominance in the commercial launches (see Chapter 3) has shrunk over the year and the International Launch Services (ILS), an American-Russian commercial company that runs the Angara and Proton rockets, only account nowadays for 10% of the global market share. There are plans of expansion, however; in 2016, Roscosmos opened a new launch complex, Vostochny, which will eventually take over most the ISS-used Baikonur Cosmodrome in Kazakhstan as the primary launch facility.

In the aftermath of the collapse of the Soviet Union, Russia has gone out on the market to strike agreements and joint venture to continue its various space activities, some of which have been going on for a while with mixed results.

The following are only a few among many signed by the Russian companies and by the Russian space agency in the last twenty years (in addition to the most famous, the ISS, examined in the case study in Chapter 6):

A. Sea Launch consortium[37]—US, Russian, Ukrainian and Norwegian companies. It was established back in 1995 and launched its first rocket in 1999. After being put in hiatus in 2014 (due to the Russian intervention in Ukraine), the consortium eventually found a buyer in 2016 (namely S7 Group, Russia's biggest domestic airline), which intends to resume operations.
B. Starsem[38]—French/European and Russian operators. Starsem was created in 1996 to commercialise the Soyuz rockets at an international level, and it lists among its main shareholders Arianespace and Roscosmos.
C. The German-based Eurockot Launch Services GmbH. The company is a commercial spacecraft launch provider, founded in 1995 as a joint venture of ArianeGroup and Khrunichev Space Centre. Operating from the Plesetsk Cosmodrome, Eurockot launched satellites[39] into LEO for institutional and commercial satellite operators. "Eurockot has become one of the leading launch providers for climate research, remote sensing and science satellites requiring high inclination and sun-synchronous orbits" (Eurockot 2019, online). The last launch occurred in April 2018; since then, activities have been in redefinition.
D. Lockheed Martin, which combined forces with Energia and Khrunichev Space Centre through ILS.

Roscosmos has a working relationship with ESA, and many ESA missions have been launched using Russian rockets. An example was ExoMars in 2016, the result of cooperation between the two space agencies, launched by the Russian Proton. There are many ongoing initiatives on paper for Roscosmos, first of all, revamping optical satellites, setting up a new space station after the demise of ISS (to which Russia is the second largest investor) and landing a probe on Mars surface.

Case Study: Brexit and the Future of the EU Space Industry

A lot has been written about Galileo, the €10 billion, all-European satellite navigation system, after the referendum held in the UK in June 2016, whose outcome started the process called "Brexit" and that will perhaps result, if nothing is done to reverse its course, in the exit of the UK from the EU originally planned for 29 March 2019 and then delayed to a still uncertain date at the time of writing this case study (but not later than 31 October 2019; *FT*, 12 April 2019).

Among the numerous difficulties of disentangling the UK from more than forty years of EU integration, there is the unique challenge it poses to transnational projects with a security component.

Galileo represents here a textbook case.

The European navigation system is to date partially operational but not completed yet; when it will be, in the mid-2020s, Galileo will surpass any existing geo-localisation services in terms of accuracy and precision, the US GPS included. It is its technical prowess that is creating political issues, and more specifically, the feature called PRS (i.e. Public Regulated Service). The PRS makes Galileo a secure, encrypted navigation system highly resilient to external threats, designed, as it is, to work even when the other navigation systems are jammed. It is therefore deemed essential as a vital back-up system for both civil and military purposes.

The UK has been involved since day one in the design and the construction of Galileo with Surrey Satellite Technology Ltd (see Sect. 4.4), the company that manufactures the payload of Galileo's satellites. This is where the problem started: for security reasons, the EU does not allow a non-member state to participate in the manufacturing of a sensitive, security-essential components, albeit a non-member access to the PRS has been negotiated with non-EU, ESA members such as Norway and Switzerland (ESA 2010, 2018a), and talks are ongoing with the USA. But none of the non-EU countries are involved in the design and handling of Galileo's components, as in the case of the UK.

The row around Galileo's membership and manufacturing contracts began in January 2018 when the supplier contracts for Galileo had to be renewed. As the renewal bidding procedure began, the UK was threatened with exclusion due to Brexit negotiations (Reuters, 14 May 2018). The UK reaction was strong-worded. The UK Defence Secretary Gavin Williamson declared in March 2018 that "the government has been clear that we want our critical role in this important project, which will help strengthen European security, to continue as we develop our deep and special partnership with the EU. This could only happen with complete UK involvement in all aspects of Galileo, including the key secure elements which the UK has unique specialisms in and have helped to design and implement" (The Register, 10 March 2018).

The dispute did not end at payload manufacturing, but also affected, among other things, the PRS ground services, managed by Airbus in Portsmouth (UK). "The GSMC has its main and operational location in France and its back-up site in the United Kingdom (Swanwick). As a consequence of the United Kingdom's withdrawal from the EU, the GSMC's back-up site needs to be relocated from the United Kingdom to one of the 27 EU Member States" (The EU press release, 2018). Spain seemed the most likely replacement.

After those initial exchanges, the UK threatened to retire from the project (*FT*, 2 May 2017), build an all-UK navigation system (UKSA, 2 May 2018) and even prevent British companies from taking any part, sensitive technology or not, in Galileo (BBC, 16 May 2018). Theresa May's government also warned the EU that the UK could claim back £1.2 billion spent for the development of Galileo (*The Independent*, 19 May 2018).

There are many issues surrounding this complex affair, first of all regarding the ownership and the management of Galileo itself. Galileo is an EU project, but ESA has been involved in it since the beginning, at a financial level as well.

More precisely, definition, development and in-orbit validation were carried out by ESA and co-funded by ESA and the EU. This initial phase created a mini-constellation of four satellites and a reduced ground segment established to validate the overall concept, ahead of further deployments. The success of phase one led to the current Full Operational Capability phase, fully funded by the EU and managed by the EU Commission. However, the EU and ESA have a delegation agreement by which ESA acts as the design and procurement agent on behalf of the EU Commission.

While the EU's position has been so far rather clear in the sense of excluding the UK from the manufacturing of Galileo as a result of Brexit, ESA took a different stance. ESA's Director Jan Woerner did not elaborate on Brexit, but he spoke about the UK in space. "The UK is part of ESA. ESA is not part of the EU. There is no change in UK participation in ESA as a result of Brexit" (*EuObserver*, 2018, online). Moreover, the space agency has warned that any change in the manufacturing of Galileo's components will likely results in a delay to get the system fully operational.

As of December 2018, it seems likely that the UK will indeed manufacture its own navigation system. The government acknowledged the exclusion from Galileo and declared its intention to pursue a UK-based navigation system, in the words of the Prime Minister Theresa May on 30 November 2018: "I have been clear from the outset that the UK will remain firmly committed to Europe's collective security after Brexit. But given the Commission's decision to bar the UK from being fully involved in developing all aspects of Galileo, it is only right that we find alternatives" (*The Telegraph*, 1 December 2018).

The government has already earmarked £92 million (out of the £3 billion Brexit readiness fund) for an 18-month feasibility study on the design and development of a domestic alternative, which independent experts put at a global cost in between £3 and £5 billion. The leading part in the study is going to be the UKSA, with the cooperation of the Ministry of Defence (Uk.gov press release, 1 November 2018).

However, this will not mean an estrangement of the UK from the European space activities; the contrary is true. As already stated in the UKSA's recent Report, the UK increased its contribution in ESA to E 334.8 million, making it the fourth largest contributing country.

And, with the lack of overall clarity that still abounds about Brexit, it is too soon to know if this is the way Galileo Affair will truly end.

List of References

Amos, J. (2018). *UK ups the ante on Galileo sat-nav project.* Available from http://www.bbc.com/news/science-environment-44116085. Accessed 24 May 2018.

BBC News. (2018, August 28). *Money Released for 'UK sat-nav' Study.* Retrieved 14 November 2018, from https://www.bbc.com/news/science-environment-45336001.

Busby, M. (2018). *UK Plans Own Space Programme After Dispute with EU Over Galileo Project, Defence Secretary Announces.* Available from https://www.independent.co.uk/news/uk/home-news/brexit-galileo-satellite-programme-space-dispute-williamson-barnier-clarke-a8360276.html. Accessed 14 September 2018.

ESA. (2010). *Norway Signs Galileo Agreement with Commission.* Available at https://www.esa.int/Our_Activities/Navigation/Norway_signs_Galileo_agreement_with_Commission. Accessed 24 May 2018.

Financial Times. (2018). UK Makes Galileo Satellite a Condition for EU Defence Collaboration. Available from https://www.ft.com/content/e8d45d26-5f34-11e8-9334-2218e7146b04. Accessed 24 May 2018.

Financial Times. (2019). Brexit Timeline: Key Dates in the UK's Divorce from the EU. *Financial Times* [Online]. Available at https://www.ft.com/content/64e7f218-4ad4-11e7-919a-1e14ce4af89b. Accessed 12 April 2019.

GOV.UK. (2018, August 29). *Space Sector to Benefit from Multi-million Pound Work on UK Alternative to Galileo.* Press Release. Retrieved 17 December 2018, from https://www.gov.uk/government/news/space-sector-to-benefit-from-multi-million-pound-work-on-uk-alternative-to-galileo.

Paladini, S. (2018). *Brexit and the EU Space Industry.* https://centreforbrexitstudiesblog.wordpress.com/2018/05/29/brexit-and-the-eu-space-industry-the-galileo-conundrum/.

The Telegraph. (2018, December 1). What Is Galileo and Why Is Britain Set to Build a Rival Satellite System? Retrieved 17 December 2018, from https://www.telegraph.co.uk/news/2018/12/01/galileo-britain-set-build-rival-satellite-system/.

UK Space Agency. (2018). *Prosperity from Space: A Partnership Strategy for UK, Space Growth Partnership*. Available from http://www.ukspace.org/wp-content/uploads/2018/05/Prosperity-from-Space-strategy_2May2018.pdf. Accessed 24 May 2018.

Notes

1. At present, the space sector is governed by the following EU legislation:

   A. Regulation (EU) No 1285/2013 on the implementation and exploitation of the European satellite navigation systems, Galileo and EGNOS;
   B. Regulation (EU) No 377/2014 establishing the Copernicus programme;
   C. Decision No 541/2014/EU establishing a framework for space surveillance and tracking support (SST);
   D. Regulation (EU) No 912/2010 setting up the European GNSS Agency;
   E. Decision No 1104/2011/EU on the rules for access to the public regulated service provided by the Galileo system;
   F. Council Decision 2014/496/CFSP on aspects of the deployment, operation and use of the Galileo system affecting the security of the European Union.

   The EU Commission is the responsible agency for the EU space policy, setting its priorities and taking operational decisions. For more about the complex relationship between the EU Commission and ESA, see Verheugen (2005), von der Dunk (2003).
2. The Vice-President of the EU Commission, Maroš Šefčovič, declared in that occasion: "EU investment in space has already delivered world-class results to the benefit of European citizens and businesses. Over 10% of the EU's GDP is already dependent on space related services and major investments by the EU have enabled progress that no Member State could have achieved on its own. But we need to up our game. Space data can help our industries lead on the Internet of Things and automated driving,

and help us more accurately monitor greenhouse gas emissions to make our climate action more effective than ever before" (EU 2018, online).

3. Due to the endless controversies the Galileo case has generated since the Brexit referendum in the UK in 2016, it has been analysed in a case study at the end of this chapter.
4. The tone of the official communications also changed. Any textual content analysis is beyond the scope of this chapter, but even a summary reading can demonstrate this shift in perspective.
5. For the sake of comparison, its value was estimated at EUR 46–54 billion in 2014, representing around 21% of the value of the global space sector (PwC 2016: 5).
6. Ariane 5 is the heavy rocket. Vega is instead the light launcher, with a typical mass of liftoff of 137 t., developed by the Italian Avio Spa. It is important to notice that Arianespace sales (€945 million) are considered as domestic, even though there is a significant share of non-European customers.
7. Since the beginning, Ariane programme and ESA have been linked, as the programme was included as optional under Art. V(1)(b), Convention for the Establishment of a European Space Agency (ESA Convention), Paris, done 30 May 1975, entered into force 30 October 1980.
8. The overall statistics for Ariane rockets are impressive, with 245 launches (1979–2018), 233 of which were successful, for an overall success rate of 95%.
9. Ariane 6 can be reconfigured to cater the needs to smaller satellites, while Vega C is more powerful of the original Vega, and it can "eat up" market share from Ariane. This is not going to be an issue in Europe—where ESA is going to supervise the choice of launchers for the institutional market—but the potential for the conflict overseas do exist.
10. ESA has provided a few details about their ongoing development, however. "In February [2018], a full-scale model of the new P120C solid rocket motor was filled with 142 tonnes of inert propellant as part of this development work. Two or four will be strapped to Ariane 6 as boosters for liftoff and a single P120C will be the first-stage motor of Vega-C" (ESA 2019: 14).
11. Note: International bodies like ESA are not included in this ranking. Moreover, the figures reported here come from government budgets, and they're not reflecting the overall investment in the space sector. For instance, China has a quite different organisation of its domestic space industry, and it is not comparable with Western models.

12. Italy has the fourth largest workforce in Europe active in the space sector, after France, Germany and the UK, even though the numbers of the Italian and the British space sector are close, with a shift in terms of ranking over the years. For more about the Italian space sector in an historical and comparative perspective, see De Maria et al. (2003).
13. However, as of December, the reform is still not over and the former ASI's chief, who was replaced for unknown reason over the year, has not been yet replaced.
14. There are some notable exceptions, such as PT Scientists and AlphaLink, but the start-up outlook is not at the level it could be expected from a country with an economy as strong as Germany.
15. For ESA's institutional design: http://www.esa.int/About_Us/Welcome_to_ESA/What_is_ESA. For details of more than 30 years of European space activities, see: http://www.esa.int/ESA/Our_Missions.
16. While it might not be immediately evident by reading the EU regulations and press release, ESA enjoys considerable autonomy and has, at times, been assuming divergent positions from the EU Commission. This is fully apparent in the Galileo case, examined in the case study at the end of the chapter.
17. "Budget allocations for the exploration of the solar system as well as for human spaceflight have stagnated ... While ESA's overall budget has vastly increased over the past eight years, it appears as though policy-makers are interested in space applications, rather than in space exploration" (Koepping 2016: 34).
18. Funding mechanisms of those programmes are quite complex. Galileo's details have been given in the case study. For Copernicus, see the official website at: https://ec.europa.eu/growth/sectors/space/copernicus_en.
19. Like Galileo, Copernicus is a European venture. For instance, the most recent satellite launched for the Copernicus programme was the Sentinel 5P, an air-pollution monitoring satellite built by Airbus in Stevenage, UK. For updates about Copernicus features, see: https://www.esa.int/Our_Activities/Observing_the_Earth/Copernicus.
20. Hollinger, Peggy, 'UK given ultimatum on Copernicus space project,' *FT*, 20 July 2017, https://www.ft.com/content/a7f188b0-6d68-11e7-b9c7-15af748b60d0.
21. The article continues with the following points "a) by elaborating and implementing a long-term European space policy, by recommending space objectives to the Member States, and by concerting the policies of the Member States with respect to other national and international organisa-

tions and institutions; b) by elaborating and implementing activities and programmes in the space field; c) by coordinating the European space programme and national programmes, and by integrating the latter progressively and as completely as possible into the European space programme, in particular as regards the development of applications satellites; d) by elaborating and implementing the industrial policy appropriate to its programme and by recommending a coherent industrial policy to the Member States" (Art. 2, ESA).

22. This also enables ESA to share the ESA classified information it collects with third parties such as the EU Council (supported by a set of dedicated Security Agreements) (ESA 2014).
23. "Ensure the protection of ESA's present and future assets; Identify capability gaps in coordination with the relevant stakeholders (national or European institutions, private operators); systematically consider dual-use requirements from early phases of programmes as well as in the evolution of flagship programmes; Identify niches (services, technologies and missions) in support of European industry competitiveness; Propose new activities with appropriate governance models in coordination with Member States and other stakeholders; Contribute to maximising the use of existing space assets for security and develop new services based on these assets" (ESA 2016, online). Differently from other programmes, SSA is an optional ESA programme, carried out since 2009 with financial participation by 14 member states. Its mandate was extended until 2019. The second phase of the programme received €46.5 million for the 2013–2016 period (ESA 2017).
24. This is an area where the EU budget is still quite low (in the new MFF only 500 million Euro have been earmarked for it), albeit the new commitment of the EU to space just unveiled in June 2018. It is still unclear how the new equilibrium between ESA and the EU will work.
25. For a detail of the services, see: https://sst.ssa.esa.int/cwbi/general/optic_sens.xhtml and https://sst.ssa.esa.int/cwbi/general/radar_sens.xhtml.
26. ESA's contributor shares have been discussed in the former section. For what concerns more specifically the UK that contribution has been increasing over time: in December 2016, at the ESA Council of Ministers meeting, the UKSA committed over €1.4 billion to ESA for the period of 2017–2019.
27. Compared to other national space agencies, the UKSA is rather lean in terms of staff. As of 31 March 2018, it employed 153 staff across its six

directorates. The UKSA is headquartered in Swindon, England, with two other sites in Harwell (near Oxford) and London.

28. A few details about the Anglo-French company and its links to ArianeGroup have been mentioned in Sect. 4.2. Airbus is present in Fortune 500 2017 (# 105), providing about 25% of the planet's telecommunications satellites and leading some of the most complex science and exploration missions in the world.
29. A Fortune 500 company in 2017 (# 498 with US$23,592 of revenues), BAE system provides, with its space division, high-tech products and services to civil and commercial space agencies for use on Earth, near space and deep space missions. Corporate website: https://www.baesystems.com/en-uk/capability/space-systems.
30. Others are less internationally known but equally innovative, such as Earth-i (video-services), MC Clyde (cube-sat manufacturer), OneWeb (small satellites for telecom) Astroscale (debris removal).
31. A more recent report, published on January 2019 by Frost and Sullivan, gives this percentage at 73.8%.
32. For the technicalities related to small and microsatellites, their denomination and the discussion about their self-decaying orbits, see in Chapter 8.
33. By all accounts, Russian space programme started in 1957, when the Soviet Union launched the world's first artificial satellite Sputnik 1 into orbit.
34. At present, a Soyuz ST is adopted by Arianespace as the medium launcher in the Ariane fleet.
35. The latest (as of December 2018) was the successful 1895th launch from Baikonur Cosmodrome on 3 December 2018. The Soyuz MS-11 manned transport spacecraft was set on target orbit for a mission to the ISS, in its thirteenth Soyuz mission in 2018.
36. This double role of a space entity can be found also in other countries, such as India, with ISRO (see this chapter). Company data of Roscosmos have been provided in Chapter 3 case study.
37. For background about the initiative, see: https://www.energia.ru/en/launchers/launchers.html. Incidentally, the launch of spacecraft from the sea has some specific legal implications (Kerrest de Rozavel 1997).
38. Starsem's website provides precious data about Soyuz long history of launches. http://www.starsem.com/news/headlines.html.
39. The company used its Rockot launch system, based on stocks of the Russian SS-19 Intercontinental Ballistic Missiles.

References

Aerospace Technology. (2015, December 21). *The 10 Countries Most Active in Space*. Available at https://www.aerospace-technology.com/features/featurethe-10-countries-most-active-in-space-4744018/. Retrieved on 10 January 2017.

Ariane Group. (2017). *Ariane Group HomePage* [Online]. Available at https://www.ariane.group/en/. Retrieved on 28 February 2018.

A.S.I.—Agenzia Spaziale Italiana. (2018). *Lo spazio ha una nuova legge* [Online]. Available at https://www.asi.it/it/news/lo-spazio-ha-una-nuova-legge. Retrieved on 28 February 2018.

Baudin, C. (1999). A Brief History of the Ariane Programme and Its Evolving Legal Infrastructure. *Air and Space Law, 24*(2), 62–69.

CNES. (2017). *Annual Report*. Available at https://www.cnes.fr. Retrieved on 19 January 2019.

De Maria, M., Orlando, L., & Pigliacelli, F. (2003). *Italy in Space 1946–1988*. Noordwijk: ESA Publishing.

De Selding, P. B. (2014, September 8). ESA Ministerial in Doubt as France, Germany Remain Far Apart on Future Launcher. *SpaceNews*. Available at http://www.spacenews.com/article/launch-report/41770esa-ministerial-in-doubt-as-france-. Retrieved on 15 January 2017.

ESA. (2005). *Cosmic Vision 2015–2025 Media Briefing. 19 April* [Online]. Available at https://www.esa.int/Our_Activities/Space_Science/Cosmic_Vision_2015-2025_media_briefing_-_19_April_2005/ (print). Retrieved on 28 February 2018.

ESA. (2014). *Law at ESA*. Available at http://www.esa.int/About_Us/Law_at_ESA/Ministerial_Council_Meetings2. Retrieved on 21 September 2018.

ESA. (2015). *Organs and Functions*. Available at http://www.esa.int/About_Us/Law_at_ESA/ESA_s_organs_and_functioning. Retrieved on 21 September 2018.

ESA. (2016, October 6). *Elements of ESA's Policy on Space and Security* (ref. ESA/C (2016)106). Available at http://www.esa.int. Retrieved on 21 September 2018.

ESA. (2017). *ESA Member States Strategies and Plans: Countries Overview*. Paris: European Space Agency.

ESA. (2018a). *United Space in Europe*. Available at http://m.esa.int/ESA. Retrieved on 21 September 2018.

ESA. (2018b). *ESA Funding Page*. Available at http://www.esa.int/About_Us/Welcome_to_ESA/Funding. Retrieved on 21 September 2018.

ESA. (2019). *Highlights 2018.* Available at http://www.esa.int. Retrieved on 10 April 2019.

ESA/EU. (2016). *Joint Statement on "Shared Vision and Goals for the Future of Europe in Space"* (ref. ESA/C/WG- M(2016)19).

EU-HR. (2016). *A Global Strategy for the European Union's Foreign and Security Policy—June, 2016.* International Cooperation and Development—European Commission [Online]. Available at https://ec.europa.eu/europeaid/global-strategy-european-unions-foreign-and-security-policy-june-2016_en. Retrieved on 28 February 2018.

Eurockot. (2019). *EUROCKOT Home Page* [Online]. Available at https://www.eurockot.com/. Retrieved on 28 February 2018.

European Commission. (2011). *Towards a Space Strategy for the European Union That Benefits Its Citizens.* COM 2011/152.

European Commission. (2013). *Final EU Space Industrial Policy: Releasing the Potential for Economic Growth in the Space Sector.* COM 2013/108.

European Commission. (2016a). *Final Communication from the Commission to the European Parliament, the Council, the European Economic and Social Committee and the Committee of the Regions. Space Strategy for Europe.* COM 2016/705.

European Commission. (2016b). *Communication from the Commission to the European Parliament, the Council, the European Economic and Social Committee and the Committee of the Regions 'European Defence Action Plan'.* COM 2016/950.

Eurospace. (2018, June 14). *The State of the European Space Industry in 2017* [Online]. Available at https://eurospace.org/wp-content/uploads/2018/06/eurospace-facts-and-figures-2018-press-release-final.pdf. Retrieved on 28 February 2018.

Fenema, H. P. (1999). The International Trade in Launch Services. In F. G. Von der Dunk & F. Tronchetti (Eds.), *Handbook on Space Law* (pp. 79–86). London: Edward Elgar.

Frost & Sullivan. (2018). *Small Satellite Report* [Online]. Available at https://ww2.frost.com/event/calendar/small-satellites-creating-new-opportunities-space-industry/. Retrieved on 1 January 2019.

Gaubert, A., & Lebeau, A. (2009). Reforming European Space Governance. *Space Policy, 25,* 37–44.

Giannopapa, C., Adriaensen, M., Ntorina, A., & Schrogl, K. U. (2018). Elements of ESA's Policy on Space and Security. *Acta Astronautica, 147,* 346–349.

Horsley, N. (2005). The Arianespace Monopoly, EU Competition Law, and the Structure of Future European Launch Markets. *Air & Space Law, 30,* 87–95.

Iserland, K. (1990). Ten Years of Arianespace. *Space Policy, 6,* 341–343.

Kerrest de Rozavel, A. (1997). The Launch of Spacecraft from the Sea. In G. Lafferranderie & D. Crowther (Eds.), *Outlook on Space Law Over the Next 30 Years*. The Hague: Kluwer Law International.
Koepping, H. A. (2016). Europe's New Wilderness: The Council's Frames on Space Policy. In T. Hoerber & P. Stephenson (Eds.), *European Space Policy*. London: Routledge.
Krige, J. (1999). The Commercial Challenge to Arianespace: The TCI Affair. *Space Policy, 15*, 87–94.
Mogherini, F.—EU High Representative for Foreign Affairs and Security Policy. (2016). *Shared Vision, Common Action: A Stronger Europe*. Brussels. EU press release 10 June 2016. Available at http://europa.eu/rapid/press-release_IP-18-389_en.htm. Retrieved on 21 September 2018.
Moslinger, M. (2017). *Enhancing Efficiency in International Cooperation Between Members of the European Space Agency*. Tallinn: Tallinn University of Technology.
Mosteshar, S. (2013). The Establishment of the UK Space Agency. In P. Hulsroj, S. Pagkratis, & B. Baranes (Eds.), *Yearbook on Space Policy 2010/2011* (pp. 115–127). Vienna: Springer.
PwC. (2016). *Space: A Cross-Disciplinary Sector* [Online]. Available at https://www.pwc.fr/en/industrie/secteur-spatial.html. Retrieved on 18 February 2018.
The Register. (2018, March 26). *Declaration of UK Defence Secretary Gavin Williamson*.
Science Business. (2018, June 6). *EU Denies 'Power Grab' on European Space Agency*. Available at https://sciencebusiness.net/news/eu-denies-power-grab-european-space-agency. Retrieved on 17 December 2018.
Suzuki, K. (2003). *Policy Logics and Institutions of European Space Collaboration*. Oxford: Ashgate.
Tarasenko, M. (1996). Current Status of the Russian Space Programme. *Space Policy, 12*(1), 19–28.
UKSA. (2017). *UK Space Agency Funds Satellite Solutions for Developing Countries*. Available at https://www.gov.uk/government/news/uk-space-agency-funds-satellite-solutions-for-developing-countries. Retrieved on 11 February 2018.
UKSA. (2018). *UK Space Agency*. Available at https://www.gov.uk/government/organisations/uk-space-agency. Retrieved on 11 February 2018.
Venet, C. (2012). European Space Governance: The Outlook. *Space Policy, X*, 59–60.
Venet, C., & Baranes, B. (2012). *European Identity Through Space: Space Activities and Programmes as a Tool to Reinvigorate the European Identity*. Berlin: Springer.

Verheugen, G. (2005). Europe's Space Plans and Opportunities for Cooperation. *Space Policy, 21*(2), 93–95.
von der Dunk, F. (2003). Towards One Captain on the European Spaceship: Why the EU Should Join ESA. *Space Policy, 19*(2), 83–86.

5

Beyond Europe: Space Powers Around the World

5.1 The USA

No country currently spends, or has ever spent, as much as the USA in the world's space industry. At the moment of the Apollo missions[1] that put a man on the lunar soil, the USA had a budget allocation for space missions of almost 4.5% of their federal budget—an astronomic amount if compared to the current allocation. Moreover, if there is a name known over all the others in the space sector is NASA. Since its creation in 1958, the space agency has prominently featured in all the most daring endeavours in space, from the exploration of the Solar System to sending the man to the Moon, flying Space Shuttles, and playing a leading role in the ISS. No other space agency has polarised the world's attention over its missions over the decades.

And while its budget has been greatly reduced since the space race years—it hovers about 0.5% of the GDP—it is still the largest in the world (Table 5.1).

Albeit the last decade's turbulent times and a federal trade deficit exceptionally large, aerospace has maintained a strong performance in the USA, especially if compared to other sectors. In 2017, the sector recorded

S. Paladini, *The New Frontiers of Space*,
https://doi.org/10.1007/978-3-030-19941-8_5

Table 5.1 Total industry output breakdown (2017) (in US$ million)

End use manufacturers	447
Aircraft systems	339
Land and naval system	51
Space systems	40
Cyber	15
Supply chain	418
Total output	856

Source AIA (2018)

US$865 billion output, led by commercial aircraft and civil space systems, with a positive trade balance of US$86 billion (on total aerospace exports of US$143 billion). This figure accounted for 9% of all the US exports (services excluded; AIA 2018). On a comparative basis, the sector exports were the third largest, behind electronic products and the energy sector.

The lion's share (52%) went to the producers of end-user goods and services including aircraft, space systems, land vehicles, ships and armaments. And while the output was 1% lower than 2016 (and 2.1% down from US$883 billion in 2012), all the provisional indicators forecast 2018 stronger than the previous one.

Aerospace as a whole employed in 2017, 2.4 million people—80,800 direct jobs in the space system—two-thirds of which concentrated in eight states: California, Washington, Texas, Arizona, Connecticut, Florida, Georgia and Ohio. While California has the largest workforce, Washington is the first country exporter, with US$42.4 billion (30%).

Spaceports and their value chain of goods and services generate part of the sector revenues. For what concern the USA, commercial launches can lift off either from the two federal launches site (Cape Canaveral, Florida, and Vanderberg, California) or from one of the eight spaceports shown in the map (Table 5.2).

The oldest of them is California Spaceport, which received in licence in 1996: "Aerospace and defence-related research and development spending from both industry and government experienced relatively flat growth in 2017 compared to the previous year. On the industry side, the top 25 U.S. aerospace and defence companies allocated a disclosed total of

Table 5.2 The world's most active spaceports: selected launching sites (Number of launches from 1958 to 2018)

Russia	Plesetsk	1569
Kazakhstan	Baikonour	1396
USA	Cape Canaveral	858
USA	Vandenberg	615
USA	Wallops Virginia	34
French Guyana	Guiana Space Centre	275
India	Satish Diwan	61
China	Jiuquan	106
China	Xichang	123
China	Taiyuan	66
China	Wenchang	3
Japan	Uchinoura	31
NZ	Rocket Lab	3
Iran	Iman Khomeni	4
Israel	Palmachin	8

Sources Author's elaboration on CSIS Data (Retrieved from https://aerospace.csis.org/data/spaceports-of-the-world/ on 10 January 2019)

$16.6 billion on research and development, which represents 4.2 per cent of those companies' total revenues for 2017. Over the past five years, research and development spending from the top 25 has grown by 11 per cent in current dollars, or 2.5 per cent as measured in constant dollars. U.S. government spending on research and development attributable to aerospace and defence totalled $ 83.9 billion in 2017. Over the past five years, this spending has dropped by eight per cent in current dollars, or 16 per cent as measured in constant dollars. By agency, the U.S. Department of Defence accounted for 86 per cent, or $ 71.9 billion, of total spending, while the National Aeronautics and Space Administration accounted for the remaining 14 per cent, or $ 12 billion" (AIA 2018: 14).

Looking at the export data, the first partner in 2017 in terms of destination was China, accounting for US$16.3 billion (11.5% of total industry exports), followed by France (US$13 billion), the UK (US$10.1 billion), Canada (US$7.4 billion) and Germany (US$7.4 billion). Other important markets for the USA are Japan, UAE, Saudi Arabia, Brazil and Singapore. In terms of regions, however, Europa was the largest recipient, accounting for 36% of all the exports.

Prominent import markets are more or less the same that already figured in the export top ten list, with the notable exception of China (present only at the tenth place, with US$1.2 billion), Mexico (US$2.2 billion) and Italy (US$1.7 billion).

And while the perspectives for the sector remain overall positive for the coming years, the looming possibility of the imposition of tariffs due to trade wars (with China, just to mention one) threatens this forecast.

Disaggregating further the industry data, AIA remarked that the space sector is "on a growth curve" (AIA 2018: 5), accounting for US$41 billion of the grand total. Commercial launching and satellite manufacturing have been the two subsegments leading the growth, with 29 launches in 2017 (+7, compared to 2016) and 268 US-made spacecraft delivered to commercial and government customers (+184).

It must not come as a surprise; albeit the growing presence of other countries in space, the USA still accounts for approximately one-third of the operational spacecraft currently in orbit around the Earth, and its efforts in space research and exploration efforts, led by NASA,[2] are unmatched by the rest of the world (Table 5.3).

The launching industry—an important component of the US space sector, with a long and at times eventful history—and the rise of the private space operators, contracted by NASA, have been covered in Chapter 3.

It is the case to observe here that 1986 was a watershed year in its history for more than one reason: the Challenger explosion also put a stop to the use of the Space Shuttle for lifting commercial payload into space (mainly satellites into their chosen orbit). From that moment on until 2011—the year of the last flight—the only missions the Shuttle flew were to take astronauts to the ISS.

And, until the new Orion becomes operational, satellites will keep going into space by private companies or by NASA-contracted and private-developed rockets.

Finally, NASA, the most famous space agency in the world.

The agency is an independent body of the executive branch of the US government, and its Administrator is nominated by the US President and confirmed by the Senate's vote. It answers directly to the US President and enjoys a degree of discretionary power (Wild 2017) through the so-called Other Transaction Authority, which allows the agency to decide the

Table 5.3 Active launch site operator licences for commercial spaceflight, USA

Operator	Site	Location	Expiration
New Mexico Spaceflight Authority	Spaceport America	New Mexico	December 14, 2023
Space Florida	Cape Canaveral Spaceport/Shuttle Landing Facility	Florida	November 7, 2023
Alaska Aerospace Development Corporation	Pacific Spaceport Complex Alaska	Alaska	September 23, 2023
Adams County	Colorado Air and Space Port	Colorado	August 16, 2023
Virginia Commercial Space Flight Authority	Wallops Flight Facility	Virginia	December 18, 2022
Harris Corporation	California Spaceport	California	September 18, 2021
Oklahoma Space Industry Development Authority	Burns Flat, Oklahoma	Oklahoma	June 11, 2021
Space Florida	Cape Canaveral Air Force Station	Florida	June 30, 2020
Houston Airport System	Ellington Airport	Texas	June 25, 2020
Jacksonville Aviation Authority	Cecil Field	Florida	January 10, 2020
Midland International Airport	Midland International Airport	Texas	September 14, 2019
Mojave Air & Space Port	Mojave Air & Space Port	California	June 16, 2019

Source Author elaboration on FAA data, 2019 (Retrieved from https://www.faa.gov/about/office_org/headquarters_offices/ast/ on 10 January 2019)

destination of part of its budget. The recent Consolidated Appropriations Act of 2017, passed by the United States Congress, explicitly prevents slashing non-defence discretionary spending in space to protect NASA's autonomy. "The US Congress' deliberate choice has enabled NASA to pursue long-term goals that might otherwise have failed from lost project funding due to a lack of political will or support" (NASA 2018, online).

With the discretionary power comes the transparency, too, and the full disclosure of financial data is a core policy of NASA, which routinely publishes budgets, its annual expenditures for space programmes, and details of the various private and public procurements that support the various initiatives.

Over sixty years of activity, NASA carried out extensive domestic and international collaboration with universities, the private sector, commercial operators and other space agencies around the world (Wilson 2017).

Another thing widely know is that each NASA's budget is a matter of discussion and controversy, with a reason: given its amount, every year's business plan is going to affect significantly the space sector as a whole.

NASA receives its funding from the annual federal budget passed by the United States Congress. Its fund allocations have varied a lot over the years (CSIS 2018), and so did the appraisal of the initiatives, attracting at times criticism for cutting the corners and for reforms that had gone a bit too far (e.g. Dan Goldin's approach of a "faster, better, cheaper strategy of 'doing more with less'", Lambright 2007) or for the lack of focus or clarity over its strategic objectives (Table 5.4).

As much as NASA's budget is always a source of wide-open discussions and even controversies, its programme has gone through sudden changes from one administration to another especially in terms of flagship missions.

A good example is the Moon, the target of the Apollo programme, started by Kennedy in May 1961 in response to Gagarin's historical mission and that was cut short in Nixon's era after Apollo 17 landing in December 1972.

After that, the US interest in the satellite faded away and the missions came to a halt, to be resumed only in the 1990s with Clementine (1994) and Lunar Prospector (1998).

Table 5.4 Time series of NASA budget allocations

Year	NASA fed outlay US$	Total US fed spending, US$	% federal spending	President	Party
1958	89.19	71,936	0.1	Dwight D. Eisenhower	Rep
1959	145.49	80,697	0.2	Dwight D. Eisenhower	Rep
1960	401.03	76,539	0.5	Dwight D. Eisenhower	Rep
1961	744.30	81,515	0.9	John F. Kennedy	Dem
1962	1257	106,821	1.18	John F. Kennedy	Dem
1963	2552	111,316	2.29	Lyndon B. Johnson	Dem
1964	4171	118,528	3.52	Lyndon B. Johnson	Dem
1965	5092	118,228	4.31	Lyndon B. Johnson	Dem
1966	5933	134,532	4.41	Lyndon B. Johnson	Dem
1967	5425	157,464	3.45	Lyndon B. Johnson	Dem
1968	4722	178,134	2.65	Lyndon B. Johnson	Dem
1969	4251	183,640	2.31	Richard M. Nixon	Rep
1970	3752	195,649	1.92	Richard M. Nixon	Rep
1971	3382	210,172	1.61	Richard M. Nixon	Rep
1972	3423	230,681	1.48	Richard M. Nixon	Rep
1973	3312	245,707	1.35	Richard M. Nixon	Rep
1974	3255	269,359	1.21	Gerald Ford	Rep
1975	3269	332,332	0.98	Gerald Ford	Rep
1976	3671	371,792	0.99	Gerald Ford	Rep
1976 (budget transition)	953	95,975	0.99	Gerald Ford	Rep
1977	4002	409,218	0.98	Jimmy Carter	Dem

(continued)

Table 5.4 (continued)

Year	NASA fed outlay US$	Total US fed spending, US$	% federal spending	President	Party
1978	4164	458,746	0.91	Jimmy Carter	Dem
1979	4380	504,028	0.87	Jimmy Carter	Dem
1980	4959	590,941	0.84	Jimmy Carter	Dem
1981	5537	678,241	0.82	Ronald Reagan	Rep
1982	6155	745,743	0.83	Ronald Reagan	Rep
1983	6853	808,364	0.85	Ronald Reagan	Rep
1984	7055	851,853	0.83	Ronald Reagan	Rep
1985	7251	946,396	0.77	Ronald Reagan	Rep
1986	7403	990,441	0.75	Ronald Reagan	Rep
1987	7591	1,004,083	0.76	Ronald Reagan	Rep
1988	9092	1,064,481	0.85	Ronald Reagan	Rep
1989	11,036	1,143,829	0.96	George Bush	Rep
1990	12,429	1,253,130	0.99	George Bush	Rep
1991	13,878	1,324,331	1.05	George Bush	Rep
1992	13,961	1,381,649	1.01	George Bush	Rep
1993	14,305	1,409,522	1.01	Bill Clinton	Dem
1994	13,695	1,461,907	0.94	Bill Clinton	Dem
1995	13,378	1,515,884	0.88	Bill Clinton	Dem
1996	13,881	1,560,608	0.89	Bill Clinton	Dem
1997	14,360	1,601,307	0.90	Bill Clinton	Dem
1998	14,194	1,652,685	0.86	Bill Clinton	Dem
1999	13,636	1,702,035	0.80	Bill Clinton	Dem
2000	13,428	1,789,216	0.75	Bill Clinton	Dem
2001	14,095	1,863,190	0.76	George W. Bush	Rep
2002	14,405	2,011,153	0.72	George W. Bush	Rep

(continued)

Table 5.4 (continued)

Year	NASA fed outlay US$	Total US fed spending, US$	% federal spending	President	Party
2003	14,610	2,160,117	0.68	George W. Bush	Rep
2004	15,152	2,293,006	0.66	George W. Bush	Rep
2005	15,602	2,472,205	0.63	George W. Bush	Rep
2006	15,125	2,655,435	0.57	George W. Bush	Rep
2007	15,861	2,728,940	0.58	George W. Bush	Rep
2008	17,833	2,982,881	0.60	George W. Bush	Rep
2009	19,168	3,517,681	0.54	Barack Obama	Dem
2010 est	19,123	3,720,701	0.51	Barack Obama	Dem
2011	17,863	3,833,861	0.47	Barack Obama	Dem
2012	18,953	3,754,852	0.50	Barack Obama	Dem
2013	16.87	3,915,443	–	Barack Obama	Dem
2014	17.65	4,161,230	–	Barack Obama	Dem
2015	18.01	4,385,531	–	Barack Obama	Dem
2016	19.29	–	–	Barack Obama	Dem
2017	19.03	–	–	Barack Obama	Dem
2018	18.83	–	–	Donald J. Trump	Rep
2019	19.4	–	–	Donald J. Trump	Rep
2020	19.88	–	–	Donald J. Trump	Rep
2021	20.37	–	–	Donald J. Trump	Rep

(*Source* Author's elaboration on NASA sources various years)

As a matter of fact, George H. W. Bush,[3] a supporter of space exploration and more specifically of the Moon programme, proposed the Space Exploration Initiative[4] (SEI) as early as 1989, aiming at building Space Station Freedom to send humans back to the Moon and "to stay" and ultimately to explore Mars. Bill Clinton[5] shelved the programme, but the idea lingered on.

George W. Bush, when lying down "The Vision for Space Exploration (VSE)" on 14 January 2004, in the aftermath of the Space Shuttle Columbia disaster in February 2003, made a second attempt for a presence more substantial than an orbiting probe. As the SEI proposed by his father, the VSE retained the same goals but giving them focus, a timetable and budget.

As for the implementation of VSE, NASA launched the Constellation Program (abbreviated CxP); the feasibility study and review findings were presented by in NASA Authorisation Act (2005). But the lack of additional funding, required to execute it, translated in a cancellation of the Obama administration with the NASA Authorisation Act (2010).

Obama's years saw a marked change in the US space policy. With the establishment of the Commercial Cargo Program in 2005, NASA awarded contracts to Orbital ATK and SpaceX starting from 2008, to take payload and science materials to the ISS. The president committed to the NASA's Earth Science programmes, raising its funding from US$1.2 billion to 1.9 in the 2016 budget allocation.

But, for what concerned the Moon, the Obama administration saw a shift of the agency's focus from a return to the Moon[6] in favour of an asteroid mission and, later on, a human mission to Mars—the famous "Journey to Mars"[7] with the goal to send people at some moment in the 2030s.

In 2018, with the Republicans back on the saddle, the Moon reappeared again on the agency's chart, even before Jeff Bezos had committed its Blue Origin to the Earth's Satellite (each tycoon seems to have his darlings). And even if, in December of 2017s Space Policy Directive-1 (SPD-1), the Trump administration has only changed small sections in the previous administration's space policy, those modifications have somehow reversed the course. The mandate is "to lead an innovative and sustainable program of exploration with commercial and international partners to enable human expansion across the solar system and to bring back to Earth new knowledge and opportunities. Beginning with missions beyond low-Earth orbit, the United States will lead the return of humans to the Moon for long-term exploration and utilisation, followed by human missions to Mars and other destinations" (NASA 2018: 1).

In the so-called pivot to the Moon, which aims at setting up a human presence on the cislunar space on its way to Mars. The roadmap calls for the establishment of a Gateway, "a lunar orbiting platform to host astronauts farther from Earth than ever before and forge U.S. leadership and presence in the region between the Moon and Earth" (NASA 2018: 3).[8] The initial planning gives a timeline of the mid-2020s, with a habitat of 55-m2 volume, 4 crew members on a 30-day-mission in Moon's orbit.

The 2018 NASA budget amounted at US$19.5 billion[9] (NASA 2018); the 2019 budget request allocates a large part of the total funding (US$19.9 billion), that is, US$10.5 billion, to a focused exploration campaign, $4558.8 million of which devoted to the Lunar Gateway (NASA 2018).

It is too soon to know if the Gateway will eventually get humans back to the Moon; but this renewed focus on the Earth's satellite might yield multiple benefits, last but not the least, the opportunity for an international cooperation initiative led by NASA. All major space agencies—ESA, Roscosmos, CNES among the others—have ambitious plans about the Moon, and a lunar space station can well be the successor of the ISS as the "UN in space".

The future looks bright for the space sector in the USA, but recent events, such as the growing rivalry with China, have raised concerns around the world.

For what regards China, the US defence sector has long looked at Chinese activities in space with suspicion when not open mistrust, given the close relationship between the civil and military sector in space. "Unlike the United States, China does not have distinctly separate military and civilian space programs. Chinese Communist Party (CCP) leaders provide policy guidance and authorise allocations of resources for the program, and various organisations within the People's Liberation Army (PLA) execute space policy and oversee the space research, development, and acquisition process. China's military also exercises control over the majority of China's space assets and space operations. Thus, even its ostensibly civilian projects, such as human spaceflight, directly support the development of PLA space, counterspace, and conventional capabilities" (House of Congress 2016, online). This lack of transparency is a factor set to trigger the well-known dynamics of the security dilemma and make things even worse.

Enter Donald Trump and the Republicans at the helm in 2016, which have done nothing but making the US stance towards China even more hawkish. One of the new administration's worrying initiatives is the so-called Space Force, which has raised fears of an arms race in outer space. Albeit the menacing sound, the concept itself not a novelty, as there have been talks of unifying all the satellites—currently operated separately by the Army, Navy and Air Force—under a single Command for years. And the idea itself looks reasonable in terms of streamlining of operations; a SPACECOM per se is not going to change anything in the actual capabilities of the military satellites the USA already owns. If[10] any, it will probably result in a reorganisation of the whole sector and in cost-cutting due to the rationalisation of resources (Hunter and Bowen 2018). Again, it is the security dilemma's dynamics and the perception of a threat that can make something in kind an actual danger, more than the fact itself.

Things like that don't happen in a vacuum and, in this as in other cases mentioned in this book, it is still too early to know in which way things will turn out. There is still a lot of confusion regarding the structure SPACECOM will take; there are speculations it might remain attached to the Air Force, the way the Marine Corps remained linked to the Navy. Moreover, the Congress will have to legislate and pass a bill, as it did when splitting the Air Force from the Army in 1947. Provided it comes to that, the terms and the mission is going to perform are still very much under discussion.

5.2 The Rising (Red) Star in the Space Sector: China

China might not have been among the first countries to set a satellite into space (it "only" put its Dongfanghong-1 satellite into LEO in 1970) but it launched its first rocket as early as 1960, and it became the third one to send a man into space: the Chinese LM-2F rocket flew Lt. Col. Yang Liwei as part of its Shenzhou manned spaceflight programme on 15 October 2003 (or 16, depending on the time zone).

The development of the Chinese space sector as a whole happened in a similar way of the nuclear missile programme, and therefore, partly motivated by strategic security concerns and partly by China's international prestige.[11]

> Mao committed the PRC to develop strategic capabilities (nuclear weapons and long-range rockets) on the grounds that 'nuclear weapons are such a major item that without it, people will say that you don't count'. Space developments were seen in the same light, as Mao declared, in May 1958, 'we should also manufacture satellites.' (Cheng 2015: 462)

China's launching programme started back in the 1960s, with the Long March series of rockets that are, in their various evolution, still in service today.[12] The first to be developed was the LM[13]-1 series, which launched the first Chinese satellite, soon followed by the other series. The LM-2 saw its development in the 1970s and was initially only used for LEO missions. However, the following and more advanced models (2D, 2E and 2F) were used for GEO and higher orbits.[14] China continued to work on rocket development with the model LM-3 and LM-4 (Long and Li 2015).

Prestige and international standing considerations have not changed in time, and space sector made no difference. Instead, like many other things in China, what was a purely military endeavour under Mao and disjointed from economic strategies, mutated character at the end of the Cultural Revolution and Deng Xiaoping's accession to power in 1978 (Angang and Honghua 2005).

In 1979, Deng started the reform of China's economy with the 1979 Open Door Policy that created the Special Economic Zones. For space policy, this meant a radical change of perspective: Chinese space effort had to take into consideration economic development and "concentrate on urgently needed and practical applied *satellites*" (Li 2002, online).

Science and Technology was one of China's Four Modernisations Deng wanted, and one that bore direct relevance for the space sector (Finkelstein 2005). In this perspective, the National High-Technology Research and Development Plan (*guojia gao jishu yanjiu fazhan jihua;*

国家高技术研究发 展计划), aka Plan 863, highlighted the key areas of future development.

The results are under everybody's eyes.

China owns and manages the world's second largest fleet of working spacecraft in orbit and operates several constellations of satellites in all the major subsegments—navigation, remote sensing and communication, including military surveillance satellites (Hou and Hao 2015).

Moreover, the observers agree that China's space programme has matured rapidly in the last decade even in terms of management and organisation (DoD 2018). The Chinese government set up in 2015 a "Strategic Support Force" to centralise the management of space, cyber- and electronic warfare missions, and also to build extensive ground support infrastructure.

This aspect is particularly important because the administration of space programme is different from other countries; and while China has its own space agency, instituted in 1993 under the Ministry of Industry and Information Technology, the Chinese National Space Administration (CNSA) is nothing like NASA. For instance, the level of autonomy of the CNSA, and its control over the space programme, is not comparable to NASA,[15] nor it is in terms of mission selection.

It is the two conglomerates that work with CNSA that hold real decision power.

The two state-owned conglomerates created in Deng's era, the China Aerospace Science and Technology Corporation (CASC) and China Aerospace Science and Industry Corporation (CASIC), are the prime contractor of China's space services, included in the defence sector, and they are responsible for the manufacturing of launch vehicles and satellites and as well as for the operations themselves (Lin and Shi 2006).

A CASC's controlled company, China Great Wall Industry Corporation, is China's commercial satellite and launch services provider, while it is a PLA—People's Liberation Army—subordinate company, China Satellite Launch and Tracking Control General (CLTC), which is in charge of China's space launches and tracking systems, including its launch infrastructure.

Following the opening to the market promoted by Deng, China's conglomerated began their activities during the second part of the 1980s and over the 1990s, offering its launching services to put in orbit third countries' commercial satellites. The first launch China managed was in August 1987, when the LM-2C successfully provided piggyback launch of microgravity test instrument for French Matra Marconi Company, marking the beginning of China's international cooperation in the aerospace industry.

After a successful decade, things came to a halt in 1999, when the American began imposing export control on all the satellites using American technology, the fact that instantly made the Chinese launchers less competitive.[16] It was only in 2011 that China got back to the commercial launching market, with the European communication satellite provider Eutelsat.

On the other hand, this ban encouraged the Chinese to start the production of their own satellites, in addition to launching them, to hand them over to the clients in a sort of turnkey project, the so-called delivery in orbit.

The first sale was a communication satellite to Nigeria in 2007 (for US$250 million, launch included). In 2010, another sale of the same kind, for *about $300 million, mainly financed by the China Development Bank, was made to Bolivia.*

As a result, China managed to capture about 19 and 16% of the world's commercial launching market (which not included launches for China's government and military sector) share in 2011 and 2012, albeit that share has been fluctuating over the years.

For what concerns financial figures, not a lot is known about China's official budget allocation, apart from what goes into the five-year plan: "While China does not release budget information for its space activities, its spending on space is likely growing, although still dwarfed by that of the United States. Public reports have estimated that China spends $2 billion to $6.1 billion per year on its space program, in comparison with the OECD's estimates of $39.3 billion spent by the United States and $5.3 billion by Russia in 2013" (House of Congress 2016, online).

In 2017, according to US estimates (GAO 2019), China devoted about $11 billion to the space sector, the second highest investment after the USA.

Often working in close cooperation with the state-owned enterprises, there are a few private companies that have recently entered the launching

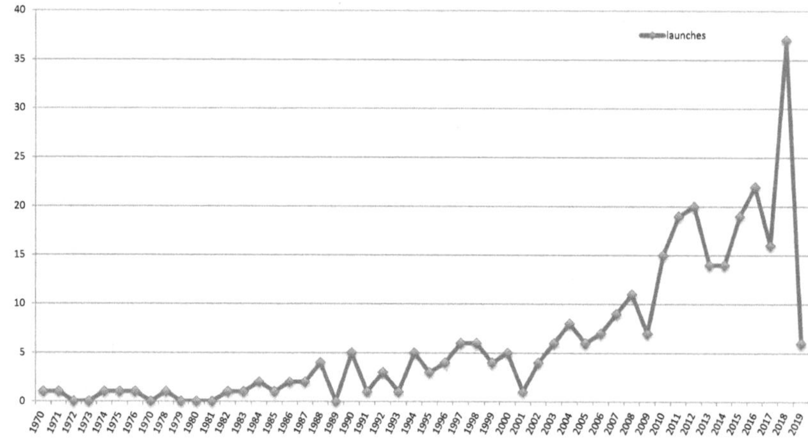

Fig. 5.1 Timeline of Chinese launches long March (up to 30 April 2019) (*Source* Author's elaboration on country's official sources)

industry and that have been highlighted in Chapter 3 when discussing the rise of the private operators. It started in 2014, when the Chinese government made the decision to allow private investments in the space sector. Landspace, a private company founded by Tsinghua University in 2015, was one of the first (Space Daily, 24 August 2018). Since then, and surprisingly it might seem, their presence grew rapidly. The country has now more than 60 private companies active in the commercial space industry, both as satellite producers and as launchers, and this number is multiplying.

The present and even more the future look bright: "Today China is a major space power with a record of successful crewed spaceflights; two space stations, with plans for a third; lunar orbiters and a lunar rover; and a program to put Chinese taikonauts on the Moon. To achieve these feats, China has an advanced family of rockets, the Long March series, that is used to launch satellites and the crewed Shenzhou spacecraft" (CSIS 2018: 6) (Fig. 5.1).

2018 has been a year of success for China, with an impressive records of rockets launch (39,[17] 34% of the world total; The Space Foundation 2019), and it will be remembered as the first year in which the Asian country managed to send more spacecraft[18] into space than any other.

But China aims further than the satellite industry (People's Daily 2018). The country runs an ambitious Lunar Exploration Programme (CLEP), which has seen the lunar rover Jade Rabbit [Yu-Tou] landing on 2014, after having entered the field of deep space exploration in October 2007 with the LM-3A successfully launching the spacecraft "Chang'e 1" to the satellite. China has now sent a new spacecraft (Chang'e 4) in December 2018, to explore the far side of the Moon.

The development of the next generation of rockets, the LM-5 and LM-8, is aimed at continuing the manned exploration of space and eventually sending probe missions to Mars: "The Long March 5 (LM5), which will be the workhorse of the new fleet, is designed to be a heavy-lift launcher having a payload capacity of 25 tons to LEO and 14 tons to GEO. LM-5 has primarily been conceived for launching the large modules for the country's future space station, but up to six variants will be developed for different purposes" (Aliberti 2018: 34).

The development of the Chinese navigation system, BeiDou, has been covered in the case study, as one of the perks of President Xi Jingping's flagship BRI initiative and the object of widespread attention around the world (Hao and Tronchetti 2017).

Albeit the impressive achievements and even the extended dearth of its international cooperation with emerging economies, there have been a few concerns in the USA about the military implications of China's space programme, which have contributed to exclude Chinese participation in the ISS (O'Brien 2013). This is why, as the USA and Russia before, also China invested into domestic space stations, "*to protect China's national rights and interests*" (PRC 2016, online).

It started with two prototypes; one, launched in 2011 and that crashed into the ocean in 2018 (Tiangong 1), the other still in orbit (Tiangong 2). The country also launched its first resupply spacecraft, Tianzhou-1 (TZ-1), to dock with and transfer fuel to Tiangong-2, simulating the Soyuz's rapid docking procedure.

An ISS-type, modular space station,[19] called CSS (Chinese Space Station)[20] is on development; Tianhe-1,[21] its core module, has been completed in 2017 and it is due to launch in 2020 (two years later than previously estimated; *China Daily*, 10 October 2018, online). The two other 20-metric-tonne modules, Wentian and Mengtian, should follow and be assembled directly in orbit later on. If, as it looks likely, the ISS will be decommissioned in 2025, China will potentially end up with the world's only active space station in Earth's orbit.

5.3 Japan and India

5.3.1 Japan

If China is now the new "space superpower" just behind the USA and Russia, there are at least two countries that need to be taken into account in what has been defined as the "undeclared Asian space race" (Brennan and Vecchi 2011): Japan and India.

Japan is a peculiar case in the space sector, for many reasons.

First of all, while it was "only" the forth country to reach space, its original trajectory has produced, together with a few failures, also a few "first" in space exploration, due to the extreme sophistication of its technology. Compared to other countries, it has engaged into space exploration a great deal, becoming one of the selected few (see Chapter 9) to reach other celestial bodies (such as the Moon and Mars). All this has been achieved with an all-domestic approach, since, apart from recent efforts mentioned below, the Japanese space sector has always privileged its own national industry to built rockets and probes and engaging less than others in cooperation initiatives.

After an early start[22] in the development of an indigenous sounding rocket in the 1950s and the establishment of Kagoshima Space Centre to test K-6 and K-8 type solid-propellant rockets, Japan launched its first satellite (Osumi) in February 1970. Other notable achievements were the launch of the world's first radio-astronomy satellite "HALCA" in 1997 and, in 2003, the world's first asteroid sample-return spacecraft "HAYABUSA" launched by M-V rocket directly into its interplanetary trajectory. The spaceship carried back samples of the asteroid "Itokawa[23]" to the Earth.

> Over 50 years period, Japanese space launch program has progressed mainly for space science and space utilisation. It is unique in the world in that the program had been promoted only for peaceful purposes. (Endo 2015: 913)

The governance of the sector has changed over the year in terms of responsible institutions. In the 1960s, the Ministry of Education's Institute of Space and Astronautical Science (ISAS) and the Science and Technology Agency's National Space Development Agency of Japan (NASDA) were in charge. They were in 2003 consolidated into a full-fledged, dedicated space agency on NASA's model, called Japan Aerospace Exploration Agency (JAXA) (Anan 2013).

Japan is a space player to be reckoned with, operating sophisticated communications, meteorological, Earth observation and astronomical observation satellites (Suzuki 2013). And while its space programme has always been strictly civilian, a recent change in the constitution has quietly added "national security" among its aims.[24] After that, Japan was allowed to add spy satellites to its fleet.

In terms of rocket technology, Japan has well-developed production and launching facilities, for both solid and liquid-propellant rockets. While solid-propellant research was the start of domestic research, the N-series (N-I and N-II), Japan's first liquid rocket booster, and the successive H-I rockets, were based on the US Delta rocket technology, even though successive evolutions came from in-house development. As of 2018,[25] Japan uses liquid-propellant H-2A, H-2B as launch vehicles from the two launch pads of Yoshinobu launch complex in Tanegashima Space Center (TNSC).

Another rocket worth mentioning is the solid launch vehicle Epsilon,[26] which is itself a technologic merveille.

> The Epsilon rocket is a solid propellant rocket suitable for a new era. The purpose of the Epsilon rocket is to launch small satellites responsively. This means that it is focused on a low-cost, user-friendly, and ultimately efficient launch system and is aiming to take advantage of the increasing need for small satellites with a compact launch system. It is based on the earlier M-V Launch Vehicle which was a multistage solid propellant rocket with the best performance in the world, discontinued in 2006. (Endo 2015: 929)

The Japanese space programme faces challenges in terms of high costs (20–30% more than the European programme), environmental challenges (limited launch windows), and limited domestic support. Some of them have already been addressed. In 2007, JAXA privatised the H-2A rocket programme and transferred to Mitsubishi Heavy Industry (MHI). The company opened to the international market, starting a policy of cooperation with other space agencies and ending the 100% reliance on Japanese-produced parts.

Among the latest Japanese initiatives, there is the development of a regional navigation system, Quasi-Zenith Satellite System (QZSS), and a bid to join the space mining race.

5.3.2 India

India, the other competitor vis-à-vis to China, adopted a different strategy from both China and Japan; however, this strategy has evolved over the years in terms of approach and investment.

Compared to China, the country has adopted a policy of transparency more similar to NASA and publishes reports of its annual spending—about US$1 billion annually (Indian Space Research Organisation 2016; 2017a). And while it might seem, looking at the news, that the country has only recently tried to play a part into the space sector, its involvement gets back to decades ago. The Indian Space Research Organisation (ISRO), was created as long as fifty years ago (1969), carrying out in the 1970s a series of experimentations that led to programmes still in operation today, such as the satellite INSAT and IRS.

The agency has its main launch complex at the Satish Dhawan Space Centre, SDSC, from where development and launches are carried over.

The Indian space programme started in 1960 under Dr. Vikram A. Sarabhai (Indian Space Research Organisation 2017b), with the launch of sounding rocket from Thumba located in 1963. The 1970s saw the development of a multistage rocket system—Satellite Launch Vehicle-3 (SLV-3) with capabilities of reaching lower orbit with a payload sufficient to carry a satellite.

The first satellite was eventually launched in July 1980.

A major achievement was the sun-synchronous polar orbit (SSPO) with the indigenous developed Polar Satellite Launch Vehicle (PSLV), a vehicle with multi-mission capabilities that managed to launch missions to LEO, SSPO, and Geo Transfer Orbit (GTO), even with a payload of multiple satellites (Byrana 2015). In a successive, enhanced configuration, the rocket has successfully launched the Chandrayaan-1, the first Indian mission to Moon.

> Out of 23 launches of PSLV till the first quarter of 2013, 22 consecutive launches have been successful, thus demonstrating its reliability. It has also performed versatile missions with different types of orbits and payloads. PSLV has injected satellites with different sizes and mass into a wide range of mission-defined orbits. (Suresh 2015: 942)

India's rocket and spacecraft development continue. A lot is expected from the new GSLV (Gupta 2006) and, if the development of the Space Capsule Recovery Experiment (SRE) has been successful, reusable launch vehicles (RLV-TD) are still under study.

India successfully managed to commercialise its low(er) costs launching services to several states eager to send their own satellites into orbit. The contractual part is carried over by Antrix, the commercial arm of the ISRO.

As of October 2018, India launched 239 satellites for 28 different countries, all using the PSLV. Moreover, ISRO successfully launched a record 104 satellites by a single rocket in February 2017, three of them domestically built, crossing another milestone in its low-cost space-exploration programme (*WSJ*, 15 February 2017). In 2018, the country carried out 7 orbital launches (6% of the world's share), 6 of which by PSVL.

But India's most impressive achievement is the Mars Orbiter Mission, Mangalyaan, which inserted into the Martian orbit on 24 September 2014, making India the first nation to manage it at its first attempt (Park 2014). Even more impressive was the comparatively low cost of the mission, US$74 million, about ten times less expensive than NASA's Maven. India's Prime Minister, Narendra Modi, declared on that occasion that the whole mission cost less than the Hollywood movie, *Gravity*.

5.4 The Rest of the World

With more and more countries entering the space race including emerging economies (Harding 2013), the number of space agencies has grown over the years in parallel to the number of satellites sent to orbit as well as the development of spaceports. The year 2018 saw the creation of launching facilities from countries that had not actively participated in the past albeit being part to international treaties (New Zealand is a good example and will be covered later this section).

Not all of them equally take part in the space industry's value chain; the opposite is true, and it is fairly common for countries to specialise in one or more segments instead that "spread" themselves. Especially, upstream operations remain the domain of a restricted "club" even though some newcomers[27] have joined lately.

As reasons of space make impossible to discuss all of them, three have been selected for further analysis, generally because they either made the news in these last couple of years or because they have continuously played an active role in international initiatives in outer space.

5.4.1 Canada

Like the UK, Canada was one of the earliest space powers, with its 1962 launch of the satellite Alouette 1. Since then, the country has shed its own launching system, depending now on other vectors—American, Indian and Russian—to send its satellites into space. However, Canada remains one of the most active players in outer space, being one of the countries involved in the ISS and also operating fleets of satellites, for communication and EO.

The Canadian Space Agency (CSA), founded in 1989, is the entity in charge of all the space activities, and, even though its budget did not grow over recent years (According to official sources, in 2018 its operating budget at US$349 million), it has nonetheless managed to retain the country's cutting-edge profiles when space activities are about. Canada has a renowned programme for astronauts[28] (Chris Hadfield was the ISS Commander in 2013) and a substantial part of the technology used onboard of the space station is Canadian, as the robotic arm Canadarm 2. Canada has recently committed to funding its ISS share until at least 2024, in line with NASA and the other space agencies.

Talking about NASA, the two countries have maintained long-standing cooperation in space activities that predates the establishment of the CSA. NASA invited Canada's government to join their activities as early as 1969. CSA has also an ongoing agreement with ESA (Canadian Space Agency 2017b).

The Canadarm programme, which makes the country one of the world leaders for the robotic, resulted from that initiative, with a project jointly developed by SPAR Aerospace and Canada's National Research Council (NRC) (Canadian Space Agency 2017a). The first version of the robotic arm was deployed to space already in 1981, on the occasion of the second shuttle mission (STS-2). Afterwards, Canadarm has been used in many notable missions, from the Shuttles to retrieve satellites, and to the deployment of Hubble Space Telescope.

Among other science-intense activities, CSA actively engages in space exploration, providing equipment for Martian rovers and asteroid mission probes.[29]

5.4.2 Brazil

Brazil is one of the countries whose space programme has recently become associated with China, due to the launch in the last decades of a few EO satellites. But, as a matter of fact, the country's space programme began years before the partnership with Asian giant. As well as Canada, also Brazil started its space activities in the early 1960s, when the then-President Janio Quadros created the two institutions in charge of the country's space programme:

1. Grupo de Organizacão da Comissão Nacional de Atividades Espaciais (GOCNAE, established in 1961)
2. Conselho Nacional de Desenvolvimento Cientifico e Tecnologico (CNPQ).

In this early days, it was not China, but the USA the country Brazil sought cooperation with, obtaining technical assistance, formation, and even help for the establishment of launching facilities. NASA also contributed to the creation of Centro de Lancamento da Barreira do Inferno (CLBI) at Rio Grande do Norte, from where the first rocket was launched in 1965.

Albeit the initiative was successful (about 2000 rockets have been launched under this programme until 1985), the funding allocated to the sector remained limited. "While the country was one of the first in putting space activities as one of its development priorities, the allocated budget was tiny - only 2 per cent of NASA budget in the same period" (Pantoja and Ribeiro 2015: 954).

Over the years, the various institutions went through restructuring and mergers. Nowadays, the agency responsible for the space programme coordination is Comissão Brasileira de Atividades Espaciais (COBAE), while Departamento de Ciência Tecnologia Aeroespacial (DCTA), which absorbed previous manufacturers of rockets, IAE among them, is the leading industry (Ministerio da Ciencia, Tecnologia e Inovacao 2018).

On the other hand, Instituto Nacional de Pesquisas Espaciais (INPE), created in 1961, is the institution that leads on satellites.

At present, the space sector is in expansion, albeit "the Brazilian space industry is mainly composed of small- and medium-sized enterprises, mostly in charge of supplying parts, components, and subsystems. At National Systems of Space Activities (SINDAE), where INPE and IAE/DCTA are the institutions responsible for the design, assembly, systems integration, and testing of satellites and launch vehicles, respectively, the private companies' actions are restricted to supply some parts, components, and subsystems ordered by these two institutions" (Schmidt 2011: 210).

The fragmentation of the domestic industry explains why the government still is the leading force in the space industry, albeit the sector itself remains tiny[30] in comparison with others. This also explains another issue, i.e. the general low degree of competitiveness of the Brazilian space industry compared to other countries, such as South Korea or India, that have started later but achieved better results.

Only recently things have begun to change. Brazil reached out for new partners such as France, Germany and Ukraine; Embraer started to be involved in satellite construction; and country built new launching facilities at Alcantara, a place in Brazilian northwest region well positioned for all orbital directions.

> The Brazilian efforts to promote the commercial application of the Alcantara Launch Center started in 1996 when an agreement was signed between Infraero (Empresa Brasileira de Infraestrutura Aeroportuaria), AEB, and the Air Force. A division was created within Infraero intended to find potential customers for CLA with specific focus to Italy (Fiat Avio) and Ukraine (Cyclone). [Not all these agreements worked out. See, for instance, Nery and Gielow (2015)]. In 2001, all the space activities conducted by Infraero were transferred to AEB. (Pantoja and Ribeiro 2015: 961)

Another important feature of the Brazilian space programme is the already mentioned cooperation with China, which has been covered in the case study at the end of the chapter.

5.4.3 New Zealand

If the last decade has seen many emerging countries in the space sector—e.g. Bangladesh, UAE, South Korea—engaging especially in the satellite industry, New Zealand has instead concentrated of another, less evident subsegment: launching facilities.

This is less of surprising than it seems. Not only, as mentioned above, the country has demonstrated an interest in space since a long time, but it has recently taken a series of initiatives to support the development of the sector.

The New Zealand Space Agency was set up in 2016 with the express aim of looking after the country's commercial space industry and the development of the launching facilities, which, according to some estimates, might become in one decade one of the country's largest sector, already been worth $260 billion annually (*ABC News* 2018).

Geography helps, too; from the Mahia Peninsula, where the spaceport is located, an increasing buzz of activity is displayed by the Kiwi subsidiary of the US-incorporated Rocket Lab. With a record of 3 launches in 2018, the company is aiming for two launches a month in 2019, and, in time, for a weekly schedule. This has been facilitated by the signature of a bilateral agreement of the space agency with the USA for the access to the US launching technology.

But the country does not only aim at the upstream sector; with the establishment, with the government support, of the Centre for Space Science Technology (CSST) in Alexandra, NZ will also "design, build, launch, and operate satellites to provide niche space-based measurements not available elsewhere, host and broker satellite data, and create new products like maps of soil moisture using data from its own and other satellites" (*NZ Herald* 2019, online). Most of the satellites belong to the category of nano-satellites and CubeSats, but it is likely that in the future a more diverse array will be offered.

The country's legal framework is regulated by the new "Outer Space and High-altitude Activities Act" entered into force in 2017, which will prove essential to ensure a safe and nurturing environment to the emerging private space sector.

Case Study: China's Cooperation in the Space Sector—The BRI Initiative and South-South Space Cooperation

When it comes down to the space sector, China represents one of the most successful examples of cooperation,[31] the latest one represented by OBOR/BRI.

President XI Jingping's OBOR initiative—later on rebranded as "BRI"—has attracted extensive coverage around the world. OBOR stands for One Belt, One Road (or BRI, Belt and Road Initiative), referring with these acronyms to a new Silk Road Economic Belt and its parallel twenty-first-century Maritime Route through Southeast Asia, whose aim is to strengthen the links between China and the rest of the world. It is still too early to evaluate how the initiative went (it is still ongoing) but there have been some notable achievements, a few of them are in the space cooperation sector. In the literature, OBOR is often regarded as an alternative to the US efforts in Asia with its Trans-Pacific Partnership (TTP), even though neither of them requires an exclusive membership.[32]

China has been in business with other countries when satellites were about long before but OBOR has provided a substantial booster: countries included in OBOR have been the first in line to be offered China's new satellite navigation services Beidou.

An example is the China–Brazil cooperation. Those links existed before OBOR, and, as a matter of fact, the two countries have been worked together in space since 1986. In 2002, after 15 years of successful initiatives, the two sides agreed to sign a formal Space Cooperation Protocol, which would provide a more concrete framework for further joint space projects. This was presented to the rest of the world as a model of "South-South" space cooperation between developing countries. Among those projects, there was the series of China–Brazil Earth Research Satellite (CBERS)[33] manufactured in 1998 as an EO programme to monitoring areas of Amazonas.

> CBERS originally launched two remote sensing satellites CBERS-1 and -2 by using the Chinese Long March 4B rocket. Upon the success of the initiatives, the two partners have planned two additional satellites of the same class, CBERS-2B and CBERS-3 and 4, as the second stage of Sino-Brazilian cooperation. According to data from the CBRES official website, the programme has served more 35,000 users from more than 2,500 organisations registered as active CBERS clients, producing over 800,000 CBERS remote sensing images. The share of the two partners is about 50 per cent each. (Paladini 2016: 45)

While CBERS was the flagship programme of China–Brazil Cooperation, more have been going on. Another satellite initiative, Ziyuan-1, the first China–Brazil Earth Resources Satellite that was jointly developed by China and Brazil, was successfully launched in 1999. A Ziyuan-3 satellite followed on 9 January 2012.

Brazil was not the only country that has benefited from space programme initiatives with China, before and during OBOR. Other cooperation projects have seen assistance to other Latin American countries that have no in-house launching capability to send their satellites into orbit.

In another cooperation started back in the 2000s, China and Venezuela signed a memorandum of understanding on space technology and established a special joint subcommittee. Following this agreement, China began manufacturing satellites for Venezuela and helped the country build domestically designed satellites. One of the results was the launch of Venesat-1, Venezuela's first remote sensing satellite.

Other Latin American countries to which China provided assistance in space matters were Argentina, Bolivia and Ecuador.

Another good example that predates OBOR, this time in Asia-Pacific, was the initiative to creates an Asia-Pacific Space Cooperation Organisation (APSCO), put forward by China, Pakistan and Thailand in 1992 and formally launched in 2008 together with Turkey, Bangladesh, Indonesia, Iran, Mongolia and Peru. As an incentive, the signatory countries received the donation of the Chinese FENGYUNCast receiving stations when the system was established[34] in the 2000s.

What is notable about OBOR/BRI is the sheer scale of the initiative and the amount of the investment committed, which include a few aimed at cooperation in space information and technology for all countries partnering with China.

One is the Belt and Road Space Information Corridor, which is a joint initiative by SASTIND and NDRC.[35] As described on the official website, the Corridor is "a spatial information service system involving remote sensing, transmitting, knowledge and use, based on the resources of communications satellites, navigation satellites and remote sensing satellites in orbit and under construction, supplemented with space-based resources and ground information-sharing network, and designed to provide space information service and interconnectivity for countries and regions along the Belt and Road" (OBOR Information Corridor 2019, online). Among the many features of the Corridor, a *"public serve cloud platform for Chinese aerospace industry"*.

But the most important initiative China promoted to its OBOR country partners concerns the new Beidou[36] navigation satellite system, one of the most ambitious programmes[37] China has ever undertaken in the space sector, presented in a government policy paper in December 2016 (Kong 2017).

In this paper, China declared the intention to grant privileged access, in the next five years, to Beidou to all the countries that were included in OBOR/BRI (mainly East Africa and South Pacific countries). "China[38] has already shipped more than 70 million Beidou systems, which include microchips and modules, domestically and to over 90 countries, and the system is being well-received in Russia, Pakistan, Thailand, Indonesia and Kuwait" (*SCMP* 2018, online).

Beidou is still not completed, but already operational and, as of January 2019, providing regional coverage with its 23 satellites currently in orbit. It is projected to achieve global coverage by 2020, when all the 35 satellites[39] will be in place. Managed by China Satellite Navigation Management Office, it represents an important part in the overall strategy of China's "soft power". Completed or not, it is still reasonably accurate, offering a worldwide location service with an accuracy of 5 metres[40] within the Asia-Pacific region, and 10 metres in other parts of the world.

As much as cars sold in Europe now mount Galileo on their embedded navigation systems, smartphones manufactured in China, such as Huawei, Xiaomi and OnePlus, are all Beidou-compatible. According to some early estimates, the system is going to generate about US$60 billion in revenues when fully operational (*WSJ* 2016).

OBOR/BRI is still ongoing, but it is not unfair to rate it as a success, and one of the telling examples of China's cooperation in the space sector.

As for 2018, China has signed 121 agreements with 37 countries and four international organisations.

References

Belt and Road Space Information Corridor. http://tbar.casicloud.com/way/platform/training/index.ht?category=01.

Brookings. (2015, July 15). *China's Rise as a Regional and Global Power: The AIIB and the 'One Belt, One Road'*. https://www.brookings.edu/research/chinas-rise-as-a-regional-and-global-power-the-aiib-and-the-one-belt-one-road/.

China's Rival to GPS Rolls Out Global Coverage Ahead of Schedule. (2018, December 29). *South China Morning Post*. Retrieved 22 January 2019, from https://www.scmp.com/tech/big-tech/article/2179802/china-rolls-out-global-coverage-its-home-grown-beidou-satellite.

How China's GPS 'Rival' Beidou Is Plotting to Go Global. (2019). *BBC News*, 20 September 2018. Retrieved 22 January 2019, from https://www.bbc.com/news/technology-45471959.

Paladini, S. (2016). *Asian Worlds in Latin America*. London: Routledge.

The Register. (2013). *China, Brazil Mourn Loss of Their $250 m CBERS-3 Satellite After Fiery Death*. 10 December 2018. https://www.theregister.co.uk/2013/12/10/china_and_brazil_mourn_lost_cbers3_satellite/.

WSJ. (2016, December 28). *China's 'One Belt, One Road' Takes to Space*. https://blogs.wsj.com/chinarealtime/2016/12/28/chinas-one-belt-one-road-takes-to-space/.

Notes

1. Just to give a figure, NASA gave a final figure of the cost of the Apollo mission at US$25.4 billion in 1973 dollars (NASA 2014), which translates in US$2018 dollar 144.18 million (the figure comes from the Bureau of Labour Statistics consumer price index, according to which prices in 2018 are 467.65% higher than prices in 1973).
2. On the other hand, NASA does NOT operate or regulates commercial space industry, which is instead the precinct of the FAA regulatory body. As it stands today, commercial launches in the US territory must secure a licence from FAA (while government and education organisations can launch small rockets under the Amateur Rocket Activities; FAA 2018).
3. Many believe George H. W. Bush's legacy in space exploration is underestimated, and not just for the SEI. More importantly, Bush nominated

Goldin as NASA's administrator, who was instrumental for the support the Clinton administration to the ISS. "Bush's choice to replace Truly was Dan Goldin, who became NASA's longest serving administrator, staying on through the Clinton administration. Characterised as one of the most influential administrators in NASA history, Goldin took on the job of finding more support for the space station. He convinced Clinton that it could be useful in foreign policy. As a result, Clinton used the space station as a tool to ease Russia's transition to a democratic state. The International Space Station was launched in 1998 due in large part to the support from the Bush administration" (Cobb 2018, online).

4. See President George H. W. Bush's Remarks on the 20th Anniversary of the Apollo 11 Moon Landing: the speech was given on the steps of the National Air and Space Museum on 4 October 1989, and it is considered the beginning of the Space Exploration Initiative. "The time has come to look beyond brief encounters. We must commit ourselves anew to a sustained program of manned exploration of the solar system and, yes, the permanent settlement of space" (The whole speech is available at https://bush41library.tamu.edu/?id=712&year=1989&month=all).
5. Albeit the cancellation of the programme, the Clinton Administration actually had a positive effect on the space sector, further promoting commercial operation and engaging more prominently into a public/private partnership in the whole sector (Brennan and Vecchi 2011).
6. In his speech at a conference on 15 April 2010, at the Kennedy Space Centre in Florida, Obama made an important statement on this point. "Now, I understand that some believe that we should attempt a return to the surface of the Moon first, as previously planned. But I just have to say pretty bluntly here: We've been there before. Buzz has been there. There's a lot more of space to explore, and a lot more to learn when we do. So I believe it's more important to ramp up our capabilities to reach -- and operate at -- a series of increasingly demanding targets, while advancing our technological capabilities with each step forward" (Obama 2010, official transcript at https://www.nasa.gov/news/media/trans/obama_ksc_trans.html).
7. There are many who criticised the Obama administration's space policy for not keeping up the international profile of the agency in terms of cooperation. One of the examples frequently quoted is the ExoMars joint-mission planned with ESA in 2013. Originally, NASA had to provide a launch vehicle for the mission, but then, due to budget constraints, NASA backed off. As a result, ESA turned to Roscosmos.

8. "The lunar orbiting platform will host astronauts farther from Earth than ever before and forge U.S. leadership and presence in the region between the Moon and Earth. An advancement in space technology and human life support systems, the platform will offer astronauts longer stays on the lunar surface, easier crew returns, safe haven in the event of an emergency, and the ability to navigate to different orbits around the Moon. NASA's leadership is shifting from LEO and its international partnerships on the ISS, to the Moon and the creation of new opportunities that enable the National Space Exploration Campaign" (NASA 2018: 25).
9. This is what eventually the Congress granted to the agency, a far better outcome than the White House's original proposal, which would have cut the budget by 1% (about $19.1 billion). Read the story: http://www.planetary.org/blogs/casey-dreier/2018/20180322-fy18-omnibus.html.
10. As observed by Hunter and Bowen (2018, online), "the US military alone is already the world's single largest operator of satellites, with 159 – more than all Russia's military and non-military satellites combined. But while these military satellites are high-tech, they're not exactly the stuff of science fiction. Rather than carrying death rays and space marines, they are built to provide secure communications, early warning of a nuclear attack, reconnaissance, and precise navigation (among many other things)". Nor the SPACECOM itself violates the 1967 Outer Space Treaty, incidentally. See the rest of Hunter and Bowen's article for a more detailed analysis of this point.
11. Mao wanted the creation of The Fifth Academy of the National Defence Ministry (国防部第五研究院) was founded on 8 October 1956, to develop the first ballistic missile and start the Chinese space programme.
12. They are also quite reliable; according to the official statistics, LM completed 125 of 128 launches with a 98% success rate in between 2001 and 2013, in line with European and American launchers. One particular model, the LM-3A, accounted for more than 50 launches with a high success rate.
13. Reading the specialised literature on rocket launches can create confusion for the acronyms, as in many cases the indicated as CZ. The reason is simple and related to the fact that the Chinese name of the rocket is Chang Zheng 长征, which translate as "Long March". Therefore, the rockets are abbreviated as LM—for export and CZ—in the domestic market.
14. In 1984, China launched the first geosynchronous Earth orbit (GEO) communications satellite, DFH-2.

15. "The CNSA appears to be, in essence, a clearing house carrying out only a few tasks, namely, serving as the public international face of China's space programme and, second, acting as the liaison office between SASTIND and the aerospace industries. It should be recalled that CNSA was established in 1993 along with the China Aerospace Corporation (CAC) to replace the dismantled Ministry of Aerospace Industry. The underlying intention was to provide the country's space programme with a visible governmental face and apparently to separate space-related governmental functions (theoretically to be assigned to the CNSA) from industrial ones (assigned to CAC). [...] while CNSA can be seen as China's external space policy organisation, carrying out China's international obligations and representing the country in international organisations and events (e.g. the ISECG), CAC can be seen as a more powerful internal complement, wielding real power over national space programme matters" (Aliberti 2018: 5). CAC has been later on restructured in the two conglomerates CASC and CASIC. Aliberti (2018) carries out an in-depth analysis of the evolution of Chinese space administration bodies and of their current structure.
16. It was in the aftermath of the Loral-Hughes incident (*NYT*, 24 June 1998).
17. 33 of them belonged to the LM-2,3,4; 3 were LM-11, and one, the last orbital launch of the year, was the maiden launch of CZ-2D YZ-3. Finally, two were carried out by commercial launchers, Kuaizhou-1 (ExPace, a subsidiary of CASIC), and ZhuQue-1 (by the private launcher Landspace). ExPace is also known as the recipient of a US $182 million investment by the government, which to-date is "the largest investment in a non-U.S. space start-up".
18. Full statistics of 2018 launches have been given in Chapter 3.
19. "Earlier reports said the new Chinese space station will initially be much smaller than the current International Space Station (ISS), which weighs 420 tonnes, but could be expanded for future scientific research and international cooperation" (*China Daily*, 2 March 2017).
20. The core module Tianhe has been designed and built by the China Academy of Space Technology (CAST), a subsidiary of the country's main space contractor, the China Aerospace Science and Technology Group (CASC).
21. The module has been recently unveiled at the Zhuhai Airshow in Guangdong Province in November 2018.

22. "In April, 1955, the Institute of Industrial Science, University of Tokyo, conducted an experiment to launch the 23 cm-long Pencil rocket. This was Japan's first step toward the completion of the indigenous sounding rocket upon participating in the International Geophysical Year (IGY) activities from 1957 to 1958. It also marked the start of Japan's space activities and the origin of the present Institute of Space and Astronautical Science" (JAXA official website 2018, online).
23. The asteroid was named after Japan's pioneer of the domestic launching system, Professor Itokawa from ISAS, who first made experiment using solid propellant in rocket development.
24. This has been treated in Chapter 7 when discussing national space laws. China's rise in the region and its growing interest in space has provoked a worried reaction in the neighbouring countries, Japan first.
25. Among Japanese space programmes, there are the Japanese Experiment Module (KIBO)-ISS, H-II Transfer Vehicle KOUNOTORI5 (HTV5) and H-II launch vehicle. The Japan Aerospace Exploration Agency (JAXA) controls the national aerospace research and development activities. Japan has tackled re-entry technology for more than 10 years, with the HOPE-X project which was however discontinued in 2001.
26. A domestic development, it has a great reputation at home. "The Epsilon rocket aims at achieving the innovative transportation technologies that can be equally applicable to liquid fuel rockets as well as future space transportation systems". The aims of the Epsilon rocket are to achieve a level of innovation in the field that can be equally applicable to liquid fuel rockets as well as future space transportation systems.
27. Finally, in some specific cases, due to the peculiarity of their involvement in the space sector—as, for example, in the case of Luxembourg for space mining—the country section has been covered instead where it was the most relevant.
28. This programme has won Canada a lot of kudos from the rest of the world and became an example for everybody. "In the 1990s, Canada racked up a series of astronaut milestones: first woman (Roberta Bondar, 1992), first Canadian on space station Mir (Hadfield, 1995), first Canadian to operate the Canadarm (Hadfield, 1995) and first Canadian to visit the International Space Station (Julie Payette, 1999). Since then, Canadian astronauts have gone on to do spacewalks (Hadfield was the first, in 2001) and more complex duties on the space station. This culminated in 2013 when Hadfield became the first Canadian commander of the station. Canada currently has four astronauts: Jeremy Hansen and David

Saint-Jacques (selected in 2008), and Joshua Kutryk and Jennifer Sidey (selected in 2017)" (Space.com, 25 April 2018, online).

29. Recent examples are the Alpha Particle X-ray Spectrometer on Mars Curiosity and the laser on-board of NASA's OSIRIS-REx that is now exploring Bennu (Space.com 25 April 2018).
30. In 2009, only 289 people worked in the space industry, compared to 18,000 of the aviation industry (Embraer 2012).
31. It is true that China has worked on space diplomacy as a tool to facilitate its access to raw materials or to foster cooperation at regional level to establish itself as the leading country. But this takes nothing from the good of those initiatives, which resulted in positive outcomes for the whole sector.
32. Moreover, the two initiatives are conceptually different. "I argue in this essay's third section that these different efforts are in fact complementary. The kind of infrastructure financed by the Chinese initiatives is the 'hardware' of trade and investment, necessary but not sufficient to deepen integration. TPP, on the other hand, represents the 'software' of integration, reducing trade barriers, opening up services for trade and investment, and harmonizing various regulatory barriers to trade" (Brookings 2015, online).
33. CBERS-4 was a success but CBERS-3 was released a lower altitude than expected and crashed back on Earth in 2013.
34. FENGYUNCast (the Fengyun Satellite Data Broadcasting System) was created in 2003 and managed by the China Meteorological Administration (CMA), which then distributes data so collected to the regional receiving stations, including non-Chinese satellites, such as USA and Japan (Fengyun Satellite Data Center 2012, online).
35. SASTIND is the State Administration for Science, Technology and Industry for National Defence, while NDRC is the National Development and Reform Commission.
36. The name itself comes from the Chinese name for the seven stars that make up the Big Dipper. Mentioned in the press simply as "Beidou", the correct name is actually Compass Navigation Satellite System (CNSS), or BeiDou-3 in its latest development. It precursor Beidou-2 officially started up in 2005 and was announced to the world in October 2006, becoming available to the whole Asia in 2012. The original programme was instead the defence-developed Beidou navigation system used in the 1990s, in another analogy with the US GPS, available to the general

public in Asia in 2000. Details on the official website http://en.beidou.gov.cn/.

37. Only in the period 1994–2012, China has invested about $2.57 billion on the system and committed far more for successive development. Moreover, Beidou is also one of China's "megaprojects" announced by the 2006–2020 Medium- and Long-term Plan for Science and Technology Development funded by China's Ministry of Science and Technology. In a total of 16 megaprojects, 3 other them (high-definition EO system, human spaceflight, and lunar probes) are related to the space sector, in yet another proof of the attention space is now attracting from the Chinese government.
38. In terms of domestic utilisation of the system, the Chinese Ministry of Transport gave in November 2018 these figures: 6.17 million commercial vehicles, 80,000 buses in 36 cities and 370 public service ships (Beidou official website 2018, online). China has also started the development of Beidou-3, which will adopt third-generation satellites for geolocalisation.
39. Of these 35 satellites, 5 will be geostationary orbit (GEO) and 30 in Medium Earth orbit (MEO), in order to grant a total coverage. See https://directory.eoportal.org/web/eoportal/home.
40. Out of comparison, the US GPS offers accuracy down to centimetres. However, it has been observed that the main issue for Beidou to represent a real alternative to GPS will not be accuracy, but reliability, of which the US system has been given proof over decades (*BBC* 2018).

References

ABC News. (2018). *'Space is Open for Business': NZ Start-up Successfully Launches Rocket*. Available at https://www.abc.net.au/news/2018-01-22/new-zealand-successfully-launches-first-rocket-into-space/9347886. Retrieved on 25 June 2018.

AIA. (2018). *US Aerospace & Defence 2018, Fact & Figures*.

Aliberti, M. (2018). *India in Space: Between Utility and Geopolitics*. Cham: Springer.

Anan, K. (2013). Administrative Reform of Japanese Space Policy Structures in 2012. *Space Policy*. Available at https://doi.org/10.1016/j.spacepol.2013.06.001.
Angang, H., & Honghua, M. (2005). *The Rising of Modern China Comprehensive National Power and Grand Strategy* in China National Space Administration. APSCO—Convention Signed in Beijing. Resource Document.
Brennan, L., & Vecchi, A. (2011). *The Business of Space: The Next Frontier of International Competition*. London: Palgrave Macmillan.
Byrana, N. S. (2015). The Indian Space Launch Program. In K.-U. Schrogl, P. L. Hays, J. Robinson, D. Moura, & C. Giannopapa (Eds.), *Handbook of Space Security*. New York: Springer Media.
Canadian Space Agency. (2017a). *Canadarm2, the Canadian Robotic Arm of the International Space Station*. Available at http://wwwasc-csa.gc.ca/eng/iss/canadarm2/default.asp. Retrieved on 18 December 2018.
Canadian Space Agency. (2017b). *Cooperation Agreement Between Canada and the European Space Agency (ESA)*. Available at http://www.asc-csa.gc.ca/eng/programs/esa/default.asp. Retrieved on 18 December 2018.
Cheng, D. (2015). Chinese Concepts of Space Security. In K.-U. Schrogl, P. L. Hays, J. Robinson, D. Moura, & C. Giannopapa (Eds.), *Handbook of Space Security*. New York: Springer Media.
Cobb, J. (2018, December 13). George H.W. Bush's Overlooked Legacy in Space Exploration. *The Conversation*. Available at https://theconversation.com/george-h-w-bushs-overlooked-legacy-in-space-exploration-108148. Retrieved on 20 December 2018.
CSIS. (2018). *China National Space Administration Press Release*. Available at http://www.cnsa.gov.cn/n6443408/index.html. Retrieved on 11 February 2018.
DoD. (2018). *Annual Report to Congress—Military and Security Developments Involving the People's Republic of China 2018*. Washington, DC: United States. Department of Defense.
Embraer. (2012). Available at https://embraer.com/global/en.
Endo, M. (2015). The Japanese Space Launch Program. In K.-U. Schrogl, P. L. Hays, J. Robinson, D. Moura, & C. Giannopapa (Eds.), *Handbook of Space Security*. New York: Springer Science+Business Media.
FAA. (2018). *The Annual Compendium of Commercial Space Transportation: 2018*. Retrieved 11 December 2018, from https://www.faa.gov/about/office_org/headquarters_offices/ast/media/2018_ast_compendium.pdf.
Fengyun Satellite Data Center. (2012). Available at https://satellite.nsmc.org.cn/portalsite/default.aspx.

Finkelstein, D. (2005). Thinking About the PLA's "Revolution in Doctrinal Affairs". In J. Mulvenon & D. Finkelstein (Eds.), *China's Revolution in Doctrinal Affairs* (pp. 1–28). Alexandria: CNA Corporation.

Gao. (2019). Space Acquisitions DOD Faces Significant Challenges as it Seeks to Accelerate Space Programs and Address Threats. Available at https://www.gao.gov/assets/700/697998.pdf. Retrieved on 25 April 2019.

Gupta, S. C. (2006). *Growing Rocket Systems and the Team.* Bangalore: Prism Books.

Hao, L., & Tronchetti, F. (2017). Should the Red Dragon Arise? Assessing China's Options Vis-a-Vis the Enactment of a Domestic Space Resources Utilisation Law. *Space Policy, 39–40,* 9–13.

Harding, R. C. (2013). *Space Policy in Developing Countries: The Search for Security and Development on the Final Frontier.* New York: Routledge.

Hou, S., & Hao, L. (2015). Chinese Satellite Programs: An Internal View. In K.-U. Schrogl, P. L. Hays, J. Robinson, D. Moura, & C. Giannopapa (Eds.), *Handbook of Space Security.* New York: Springer Media.

House of Congress. (2016, September 27). Testimony Before the House Space Science and Technology Committee, Subcommittee on Space Hearing on "Are We Losing the Space Race to China?" Dennis C. Shea Chairman, U.S.-China Economic and Security Review Commission. Available at https://docs.house.gov/meetings/SY/SY16/20160927/105387/HHRG-114-SY16-Wstate-SheaD-20160927.pdf. Retrieved on 18 December 2018.

Hunter, C., & Bowen, B. (2018). *Trump's Space Force isn't as New or as Dangerous as it Seems.* Available at https://www2.le.ac.uk/offices/press/think-leicester/politics-and-international-relations/2018/trump2019s-space-force-isn2019t-as-new-or-as-dangerous-as-it-seems. Retrieved on 25 February 2019.

Indian Space Research Organisation. (2016). *Organisational Structure* [Online]. Available at https://www.isro.gov.in/sites/default/files/article-files/about-isro/organisation-structure/organization_chart-aug-2016.jpg. Retrieved on 10 February 2018.

Indian Space Research Organisation. (2017a). *Vision and Mission Statements* [Online]. Available at https://www.isro.gov.in/about-isro/vision-and-mission-statements. Retrieved on 10 February 2018.

Indian Space Research Organisation. (2017b). *Dr. Vikram Ambala Sarabhai (1963–1971)* [Online]. Available at https://www.isro.gov.in/about-isro/dr-vikram-ambalal-sarabhai-1963-1971. Retrieved on 10 February 2018.

JAXA. (2018). *JAXA Home Page* [Online]. Available at http://global.jaxa.jp/projects/index.html. Retrieved on 11 February 2018.

Kong, D. (2017). Shaping a Legal Framework for China's BeiDou Navigation Satellite System. *Space Policy, 42*, 26–30.

Lambright, H. (2007). Leading Change at NASA: The Case of Dan Goldin. *Space Policy, 23*(1), 33–43.

Lazzari, E. (2017). *A Study of the Economic Potential of the Local Space Sector*. Adelaide: South Australian Space Industry Centre.

Li, D. (2002). The Characteristics and Rules of Law of Space Strategy. *China Military Science, 1*, 32–39.

Lin, B., & Shi, W. (2006). Analysis of China's Aerospace Industry Development. *China Aerospace, 8*, 18–24.

Long, L., & Li, D. (2015). The Chinese Space Launch Program. In K.-U. Schrogl, P. L. Hays, J. Robinson, D. Moura, & C. Giannopapa (Eds.), *Handbook of Space Security*. New York: Springer Media.

Ministerio da Ciencia, Tecnologia e Inovacao. (2018). *Agencia Espacial Brasileira*. Available at http://www.aeb.gov.br. Retrieved on 11 February 2018.

Munsami, V., & Offiong, E. O. (2017). Should Africa Follow the European Space Governance Model for an African Space Programme? *Space Policy*, 3–1, 123–150.

NASA. (2010). *Mission, Goals, Objectives*. Available at https://www.nasa.gov/offices/emd/home/mgo.html. Retrieved on 8 February 2018.

NASA. (2014). *Public-Private Partnerships for Space Capability Development Driving Economic Growth and NASA's Mission*. Available at https://www.nasa.gov/sites/default/files/files/NASA_Partnership_Report_LR_20140429.pdf. Retrieved on 10 February 2018.

NASA. (2017). *NASA History Overview*. Available at https://www.nasa.gov/content/nasa-history-overview. Retrieved on 8 February 2018.

NASA. (2017). *Space Act Agreements Guide*. Available at https://nodis3.gsfc.nasa.gov/NPD_attachments/N_AII_1050_001D.pdf. Retrieved on 9 February 2018.

NASA. (2018). *Financial Data Transparency*. Available at https://www.nasa.gov/open/financial-data-transparency.html. Retrieved on 9 February 2018.

NASA. (n.d.). *NASA Spinoff*. Available at https://spinoff.nasa.gov. Retrieved on 9 February 2018.

Nery, N., & Gielow, I. (2015, April 16). Brazil to Cancel Rocket Launch Agreement with Ukraine. *Folha*. Available at http://www1.folha.uol.com.br/internacional/en/scienceandhealth/2015/04/1614262-brazil-to-cancel-rocket-launch-agreement-with-ukraine.shtml. Retrieved on 10 February 2018.

O'Brien, K. (2013). China and the International Space Station. *The Newsletter, 63*(1), 10.
Office for Outer Space Affairs. (2005). *Disseminating and Developing International and National Space Law: The Latin America and Caribbean Perspective.* Vienna: United Nations.
Paladini, S. (2016). *Asian Worlds in Latin America.* London: Routledge.
Pantoja, F. M., & Ribeiro, J. (2015). Space Launch Programs in Emerging Countries: Brazil. In K.-U. Schrogl, P. L. Hays, J. Robinson, D. Moura, & C. Giannopapa (Eds.), *Handbook of Space Security.* New York: Springer Media.
Park, M. (2014). India's Spacecraft Reaches Mars Orbit…and History. *CNN,* 23 September 2014. Available at http://edition.cnn.com/2014/09/23/world/asia/mars-india-orbiter/index.html. Retrieved on 10 February 2018.
People's Daily. (2018, March 8). China's Long March Rocket Family Ready for Its Busiest Year. Available at http://en.people.cn/n3/2018/0308/c90000-9434465.html.
PRC. (2016). *White Paper on Space Activities.* Available at http://english.gov.cn/archive/white_paper/2016/12/28/content_281475527159496.htm. Retrieved on 10 October 2016.
The Space Foundation (2019). *The Space Report 2018* [Online]. Available at https://www.spacefoundation.org/. Retrieved on 27 April 2019.
The Space Race: NZs Push Into a $320 Billion Market. (2019). *NZ Herald.* Retrieved 25 June 2018, from https://www.nzherald.co.nz/business/news/article.cfm?c_id=3&objectid=12016132.
Schmidt, F. H. (2011). *Desafios e Oportunidades para uma Industria Espacial Emergente: o caso do Brasil.* Texto para Discussao do Instituto de Pesquisa Economica Aplicada 16–67, Brasılia.
Suresh, B. N. (2015). The Indian Space Launch Program. In K.-U. Schrogl, P. L. Hays, J. Robinson, D. Moura, & C. Giannopapa (Eds.), *Handbook of Space Security.* New York: Springer Media.
Suzuki, K. (2013). The Contest for Leadership in East Asia: Japanese and Chinese Approaches to Outer Space. *Space Policy, 29,* 99–106.
Suzuki, K. (2017, October 10). Satellite Gives Japan Global Edge. *The Wall Street Journal.*
Wild, F. (2017). *What Is NASA?* Available at https://www.nasa.gov/audience/forstudents/5-8/features/nasa-knows/what-is-nasa-58.html. Retrieved on 9 February 2018.
Wilson, J. (2017). *NASA Centres and Facilities.* Available at https://www.nasa.gov/about/sites/index.html. Retrieved on 7 February 2018.

WSJ. (2017, February 15). India Breaks Record for Launching Most Satellites from a Single Rocket. Available at http://blogs.wsj.com/indiarealtime/2017/02/15/india-breaks-record-for-launching-most-satellites-from-single-rocket/. Retrieved on 7 February 2018.

6

Space Diplomacy: Traditional and Non-Traditional Security Issues

6.1 Space Security or Security in Space? An Introduction

As the EU Commission stated recently in an official communication, there is the growing perception of space as a "challenged" and together "challenging" place. "Space is becoming a more contested and challenged environment. New competitors – both public and private – are emerging around the world, partly spurred by the reduced costs of developing and launching satellites. Growing threats are also emerging in space: from space debris to cyber threats or the impact of space weather" (European Commission 2016: 8).

In this sense, it is true that the official discourse has changed since the start of the space age (see next section). But it is the case to say that, rhetoric of the "sanctuary" apart, space and security have been intertwined since the beginning, and so has the use of space in a military-defence perspective. As a result, at that time as much as today, "national security interests are an inherent feature of the space industry" (Jakhu and Pelton 2017: 4).

There is abundant literature that deals with the various aspects of this complex relationship, including the way it has changed and probably

S. Paladini, *The New Frontiers of Space*,
https://doi.org/10.1007/978-3-030-19941-8_6

become even more complex as the civilian-commercial actor developed besides the military, defence-led initiatives. More than often, the two sectors have worked together, both in terms of sharing data and producing dual-use devices. As the former direct of the European Space Policy Institute (ESPI) Kai-Uwe Schrogl explains, "space and security have been inseparable since the beginning of space activities. The United States and the Soviet Union/Russia have launched, until now, some 2000 military satellites. Since the Cold War, the effective functioning of these space-based assets for observation, early warning, navigation, and communications has played an essential role for the preservation of international stability and peace. That said, today's security environment is far more complex. Security threats and challenges are both civilian and military, local and global. Space-based assets are indispensable to enable global access, monitoring, and communications for addressing a whole range of security-related concerns. Approximately 20–30% of some 1000 satellites that are currently in orbit are directly related to such security functions and are usually operated by military institutions" (Schrogl 2015: 14).

This is why, when discussing security in space, it is difficult to separate commercial, space-enabling capabilities—which are meant to be in the name of scientific progress and/or commercial profit—from launching capabilities *tout court*, which obviously raise security concerns (the required technology can also be used for ballistic missiles).

When discussing security, it is important making a distinction between traditional and non-traditional security issues. Space presents plenty of both.

Space matters are generally discussed at an international level, starting from the UN, with its Committee on the Peaceful Uses of Outer Space (COPUOS) that addresses all the relevant matters. The UN also looks after the respect of the treaty that has the most contributed to maintaining an equilibrium of forces, and preventing (so far) conflicts in outer space, the 1967 Outer Space Treaty (OST), examined in detail in Chapter 7. Here, the analysis focuses instead on the kind and level of threats to the global equilibrium since the beginning of the space age and that stem from the concept of security, the way it evolved over the decades, the growing numbers of "actors" of security and the multiplication of the objects of security (Buzan and Waever 2003[1]).

The doctrine, analysed in more detail in the following section, states that space security presents two components: one, security from space, i.e. use and activities carried out in space for security purposes on Earth, and the other, security in space, which aims at maintaining outer space according to the provisions of OST, i.e. as a safe, sustainable and stable environment for all countries.

In times, the concept of space security has increasingly expanded in this second components, leading to an evolution from what was called "a two-dimensional model of military and environmental dimensions of space security" (Sheehan 2015: 34), to a wider definition that includes not only non-state actors, e.g. private operators, but also non-traditional kind of threats originating in space, both man-made and not.[2] What is more important, as the delivery of critical services from space assets requires the constant availability of space assets themselves, security in space is also important both for security from space and for a wide range of non-security applications. According to Pellegrino and Stang, "there is thus an alignment of interests between security communities and the wider public to pursue a space security discourse for preserving the stability and sustainability of the space environment, and ensuring freedom from threats to the effective access to, and use of, outer space" (Pellegrino and Stang 2016: 2).

States remain, however, the predominant actors in outer space: therefore, the different, and at times opposite way, nation states consider their national security in space contributes to make the concept wider, nuanced and more complex. This is especially true for the one that still is predominant in outer space in terms of presence, budget and influence: the USA. And while there have been in time variations in the conception of space security over the successive US Administrations—as shown in Sect. 5.4—there is one thing that has never changed: the use of space for military purposes, alone or together more civilian, more commercial-oriented uses.

In a recent CSIS report, this aspect is discussed in the current US security framework. The US military now routinely uses space systems across the full spectrum of military operations, including hunting down terrorists in remote parts of the world as well as securing a credible nuclear deterrent. However, this space-assisted space power projection can trigger a security dilemma dynamics, since "potential adversaries have taken notice. The

U.S. military's dependence on space makes these systems a natural target for adversaries to exploit. Space is simultaneously a powerful enabler for the U.S. military and a critical vulnerability" (CSIS 2018, online).

The approaches to space security[3] are as different as many space agencies exist.

And if Spain is an example of a country where, like the USA, military concerns are predominant—the Spanish Space Agency, INTA, is a public research organisation within the Ministry of Defence (INTA 2018)—India has gone into the opposite direction, with ISRO treating the military as one of its (many) stakeholders and allowing no direct influence over the agency's governance or operations. In other cases, space achievements directly translate into enhanced (at least at a level of domestic perception) national prestige, as it was the case with the Soviet at the beginning of the space age and, more recently, China Moltz (2012). The aim of this chapter is not to illustrate the shifting and ever-evolving domain of space security, too wide and out of scope in the present work. It intends instead showing which aspects—both traditional and non-traditional security—are likely to have an impact in the space sector as a whole in the coming years, playing a fundamental role in shaping those alternative scenarios proposed in the conclusions Williamson (2006).

They will, ultimately, bear the responsibility of which one, among all the possibilities now open to the worldwide space industries, will translate into reality and which will only remain in the domain of missed opportunities.

6.2 International Relations and Space Governance

When scholars of international relations (IR) come to talk about space, their approach is rather clear, and the following quote, by Lieberman, exemplifies how the general view of space security in IR is largely based on a realist ideology.

> The analysis of international relations played out in space not only suffers a paucity of theoretical discussion due to its relatively recent appearance on the academic table, but also due to the state, and land, based assumptions

> made by most relevant frameworks. [...] Realist conceptions of international relations consider the state to be the main currency of international political interactions, they also understand sovereignty and security to be the commodities that are most clearly sought, issues that are also at the heart of space power and space politics. (Lieberman 2017: 2)

It is the realist approach in space that caused the adoption of the "high-seas" framework, which is still common in this sector and which draws from the nineteenth-century to early twentieth-century well-known models of analysis for sea power. Leissle et al. (2017), among others, have illustrated an application of Alfred Thayer Mahan's famous sea power doctrine to space: "In 'The Influence of Sea Power Upon History', Mahan reviewed the role of sea power in the emergence and growth of the British Empire. In the book's first chapter, he described the sea as a 'great highway' and 'wide common' with 'well-worn trade routes' over which men pass in all directions. He identified several narrow passages or strategic 'chokepoints,' the control of which contributed to Great Britain's command of the seas Mahan (1890). He famously listed six fundamental elements of sea power: geographical position, physical conformation, extent of territory, size of population, character of the people, and character of government" (Sempa and Francis 2018, online).

As in the case of the sea, space might be suffering from the tragedy of the commons (Hardin 1968) a scenario in which independent groups make self-interest-driven, rational choices and end depleting shared, limited resources even though it is not in anyone's long-term interest for this to happen.

A perfect equivalence sea-space is not possible, for once because most of the activities in outer space are an absolute novelty in terms of characters, extents, challenges and numbers of unknowns, too. "What is unique about space is the fact that we are dealing with infinity. Whereas the terrestrial land mass and the seas have knowable finite bounds, we literally do not know where space ends or understand the implications of infinity or how we theorise about space. [...] Space is an arena in which competition and cooperation are already set forth in terms and issues reminiscent of Earth-bound phenomena. Space power includes assumptions drawn from IR theory" (Pfalzgraff 2015: 25).

However, there is some common ground; when talking about LEO (lower-Earth orbit), which can with reason be considered an Earth space as much as the atmosphere, the application of a similar framework is justified. Dolman has suggested that "geopolitical theory developed for the Earth and its geographical setting can be transferred to outer space" with the "strategic application of new and emerging technologies within a framework of geographic, topographic, and positional knowledge" (2002: 15).

When applying, as others have done, Mahan's doctrine, Sumida (2015) notes an important point: the specific security requirements of space will impose, as much as they did for high-seas, cooperative agreements at an international level. Thus, the inherent costs of space security and activities will sooner or later require the involvements of private actors in the venture for economic returns purposes. This is what happened in the last twenty years, and the trend is growing.

This approach takes into account and joins hands with another leading IR tradition, International Political Economy (IPE) and its assumptions regarding the linkages of wealth, security and government, both in its original formulation and in Strange's critical view (1994), which puts a strong emphasis on ideology in her analysis of hegemony, structural power and the relationship between states and market.[4]

More in general, a few scholars argue about the increasing relevance of space in the human imaginary, exploring the popularisation of space under a twofold perspective. Lieberman et al. (2017) one relates to the importance of ideas and ideals in the cultural arena, and a case study in Chapter 9 has been devoted to the analysis of space in its cultural aspects. The other aspect is linked to technology, much in the way Strange used it in her analysis. It is easy to show that, how Lieberman argues (2017), space was popularised as a tool to demonstrate technological superiority during the Cold War and continues to be used for this purpose today. Lawton et al. (2000) examine instead the state-based domination in the production of knowledge and technology, i.e. considering technology as the application of knowledge *in* production.

Finally, it's also important to notice that large aspect of the parallel sea power/space implies the already mentioned, common notion of prestige and reputation bestowed to the (great) powers that are able to yield this kind of power, in a definition of it that encompasses the two aspects of ideas and technologies. If anything, the Cold War exemplifies both.

This can also be reconfigured in terms of hard and soft power (Nye 1990)—a popular, albeit controversial dichotomy that has been proposed in occasion of the "peaceful" China rise as a soft power (even though this portrait of the Asian giant has become somehow outdated). The hard power aspects of space, as strongly linked military and defence components, have been explored later on in more details. But a component of soft power, while less evident, is also present, and the cultural aspects of the space race offer a good example. The case of ESA-Rosetta mission and the conscious efforts to reach out to the young generations, in Europe and elsewhere, is another.

Having set the framework for analysis, it is the time to examine the single issues that occupy now the formulation of space policy at an international level. One of the most important questions regards the governance of space, and what an increasingly busy Earth's orbit does require. And, while for what concerns space security intended in a more traditional way, cooperation has been limited, whatever the reasons,[5] civilians and commercial agreements have increasingly designed "a cooperative space" (as defined in Chapter 2).

Moltz (2014) designed a few scenarios of space governance, from one state ruling the whole system to a supranational institution of the kind of Star Trek's United Federation of Planets. There are a lot of possibilities in between, but having a weak international organisation is worse than having none (everybody remembers the League of Nations of 1920s–1930s), and the more crowded the orbits in terms of actors (public, private, even tourist operators) and space objects of different kinds, the more important a regulatory framework becomes.

Those scenarios have been described in terms of its consequences on the development of space industries in the conclusions, and they are set on a continuum between conflict and cooperation and between one hegemonic power that leads the global space policy and a cohort of competing powers. Scholars are divided in their evaluations, though; and if there are some, such as Everett Dolman, which considers the scenarios with one hegemonic power (USA) likely, even plausible, others stress out that the evolution itself of the space industry points into a different direction.

Once private actors enter more significantly into the politic aspects of space governance, they will probably prove a moderator role into space militarisation, as much as globalisation is considered a peace-enhancing factor. Moreover, Nancy Gallagher observes that the USA can't protect all its satellites, no matter how much it increases its military expenditures (Gallagher 2015).

And if Dolman and Cooper believe that the USA could probably "cow rivals into submission", Wolter wonders what are the odds of reaching what he labels as "*pax cosmica*".

Another possibility is soft governance, which hypothesises a set of "piecemeal limited treaties" (Hertzfeld and von der Dunk 2005) and ad hoc solutions case by case, like an International Code of Conduct for Outer Space Activities for government-funded initiatives and other joint efforts initiatives, such as the International Space Exploration Coordination Group (ISECG). The problem that plagues this kind of scenarios is what already observed for the UN: lack of sanctions in case of UN treaties and of enforcement mechanisms.

Finally, in addition to these comprehensive governance agreements and enforcement mechanisms, there is also the need of an international space situational awareness (SSA) body to prevent collisions[6] and prohibition of kinetic activities—both intentional and due to hazards—against space objects and that can generate lost-lasting debris and other kinds of accident. The few existing examples in this sense are, as shown in Chapter 7, a dire warning.

6.3 The Weaponisation of Space in the Twenty-First Century

The debate of the use of weapons in space and what constitutes a "weapon" is as old as the debate about space security itself. As a matter of fact, there has been a lot of confusion—and controversies, too—about the use of weapons in space and space weapons, i.e. the development and adoption of devices whose scope is clearly oriented for suborbital or orbital conflicts, as well.

Moreover, the use of space-bound technology is not per se a novelty. As Hostbeck clarifies, "space, i.e., Earth orbit, has been used for military purposes for the last 50 years, and antisatellite (ASAT) weapons have been developed during the same time without a weaponisation of space. The path of conflict in space and through space starts and ends on the ground. However, as long as ground, conflicts make use of space-bound technology, there's always the risk that the conflict could evolve in character and space-specific weapons will be developed and deployed" (Hostbeck 2015: 970).

Since von Braun described in 1952 an orbiting space station that could be equipped with nuclear weapons (Bulkeley and Spinardi 1986), three nations have so far developed and tested space weapons in an anti-satellite configuration (ASAT): USA, USSR/Russia and China, with another, India, that has recently announced plans in this sense. As it happens, there is a direct connection between the development of ASAT capabilities and nuclear weapon programmes, and this is why the USA has started the development of Earth-to-space devices as early as the 1950s.[7] These nuclear warhead-powered devices proved, however, less than optimal, and alternatives were soon developed.[8]

No official definition of "space weapon" exists to date, and it is impossible to apply what already in place for nuclear weapons: they are not the same—for one, because there is no such thing as the first-strike capability in space; moreover, they respond to different logics (Hebert 2014).

For the sake of definition, a weapon here is intended as an agent, be it a missile, energy waves or any other object, including a weapon-carrying platform, which can impact (and damage) a target. With the evolution of technology both in weapons and in the satellite business, there is now a dearth of devices that can be used in case of conflict, both in an offensive and defensive perspective.

The doctrine on weapons discusses the position of the weapon carrier in respect to the target, and a different nomination is used whether they are both based in the same domain (be it ground, sea, air) or not. This is applicable to space as well. Therefore, if the case where the targets are the satellites in space, all the land, sea and air-based platforms will be considered as "Earth". A missile[9] (e.g. an antiballistic missile, ABM) fired from a submarine would an Earth-to-space weapon (ETS), while anything already in space would be a space-to-space weapon (STS).

Table 6.1 Intentional and unintentional threats to space

Intentional threats to space systems and services	
Space based	
	Kinetic energy weapons (ASAT)
	High-altitude nuclear weapons (EM pulses)
	Directed energy weapons (laser, microwaves)
Ground based	
	Kinetic energy weapons (ASAT)
	Physical attact to ground infrastructure
	Sabotage
Content based	
	Cyber attacks
	Jamming and Interferences
Unintentional threats to space systems and services	
Space based	
	Space weather
	Cosmic rays
	Space debris
Ground based	
	Natural disasters
	Loss of utility supplies
Content based	
	Interferences from human activities
	Weather and solar disturbances

Source Author's elaboration on Sheehan (2015)

Table 6.1[10] presents a diagram with the main "intentional threats" (= weapons) in space—albeit not distinguishing between ETS, STS, STE—together with an assessment of priorities.

A brief look at the military literature shows that the defence sector started building and designing space-enabled weapon since the onset, some of them using revolutionary concepts. ASATs are the typical kind of weapons in this context, and they come in a diverse set of fashion, e.g. physical devices and waves, energy and even radio devices. ASAT missions are the most discussed and talked about due to their relevance to Space Situational Awareness (SSA). They do not necessarily involve the destruction of a satellite, because other purposes (such as jamming communication links or disabling satellite components) can also be pursued.[11]

There are other ground-to-space weapons, aimed at other airborne targets than satellites. One famous project, now shelved, is the Tactical High-

Energy Laser (THEL),[12] which ran between 1996 and 2005. THEL was developed by a Northrop Grumman-led team of USA and Israeli contractors for the US Space and Missile Defence Command, Huntsville, Ala., and the Israeli Ministry of Defence (NG 2018), and during its testing it destroyed 46 airborne targets—7 medium, 2 heavy and 1 light rockets and missiles among them.

There are, finally, weapons that launch from space to attack a target on the ground (or at sea and in the air)—the so-called STEWs. The weapons would use satellite or some sort of orbiting platform, and they could be both physical devices, such as bombs or missiles, or energy waves, like a laser. Albeit many projects especially in the first period of the space age, not a lot has passed the conceptual phase so far, and for a fairly good reason: for the convenience of other hand-ready solutions, such as ICBMs[13] or submarine-launched ballistic missiles (SLBM). "For space-to-Earth weapons to augment or replace ICMBs/SLBMs, the space systems must reach targets otherwise unreachable, must have a shorter engagement time than missiles, and must be cheaper or in some other way be superior to the existing weapon systems. If none of these criteria are met, there is no rationale in developing space-to-Earth weapons for global ground attack" (Hostbeck 2015: 971).

Moreover, placing a satellite for weapon uses faces a series of constraints, depending on which orbit it is located and the kind of object it targets. A satellite in LOE might be able to strike a target on the ground only at periodical intervals, if any at all. For the same reasons, the case of space-to-space weapons, both located in LEO, is even more complex, since in this scenario both targets would be in movement. "The orbital characteristics of a weapon in space is a mirror of the intended use of that weapon, and since the orbital characteristics are determined on launch, there is no real gain in launching such a weapon before the intended geographical area of use has been determined" (Hostbeck 2015: 974).

The only theoretically viable alternative would be to develop a constellation of satellites with offensive capabilities, able to targets ground objectives. In similar cases, issues such as absentee ratio[14] will have to be taken into consideration and would probably make the practical usefulness of such weapons an extremely expensive solution. With all likelihood, not one to be implemented in the new future.

How the international space law and treaty deals with space weapons, however intended? A simple answer, in a matter that is all but simple, it that they allow it. "Existing laws do not provide adequate restrictions on space weapons. While they do restrict weapons of mass destruction (WMD), everything else is allowed" (Hebert 2014: 4), ASAT included.

The treaties themselves and the legal basis for the peaceful utilisation of space have been analysed in Chapter 7. It is the case to observe here that, peace principles notwithstanding, all the treaties allow the use of force in a defensive function and therefore the use of weapon in space, even though not setting them up on other celestial bodies (explicitly forbidden by OST).

There is a big difference between putting arms into space and "weaponising space", and this is where the disagreement between the USA on one side and Russia and China on the other resides. In 2014, Russia and China submitted a new draft treaty on the prevention of the Placement of Weapons in Outer Space Treaty (PPWT), calling for a ban "on placement of weapons in outer space as one of the most important instruments of strengthening global stability and equal and indivisible security for all" (VOANews 2018). The USA refused to back the treaty, on the valid ground that the treaty was not dealing with the important aspect of ETS devices, such as the ASAT weapons, which both countries have in their possession and ready for use[15].

While the recent international cooperation efforts[16] do offer hope in this sense, the most dangerous thing is not an unlikely confrontation among the big space powers: it is the damage to the system itself provoked by a space increasingly chocked and threatened by debris, as the case study at the end of Chapter 7 shows.

6.4 Non-Traditional Security Issues (NTSI) in Space: A Growing Problem

While most of the security-related aspects of space are traditional, as discussed above, a few non-traditional security issues (NTSI) exist too, included in what is generally called "the global common", in which the activities of few can affect the majority (Weeden and Chow 2012). If this is true, it then makes sense to consider the space-related NTSIs as a part of environmental security widely intended.

There is extensive literature about environmental security, as this topic has been one of the first to be identified as a non-traditional security issue (Buzan and Waever 2003; Deudney and Matthew 1999) and as a geopolitics component (Elliot 2002), highlighting how environment-related aspects affect international relations in more than one way. A series of paradigms have been identified in international relations, such as environmental security studies, which consider the environment as a source, direct or indirect, of acute conflicts. "Scarcity is regarded as the fundamental problem, together with the related aspect of the access to resources" (Paladini 2015: 312); as shown in this section, the increasingly crowded LEO can well fall under this definition, even though there has been to date no conflict about its appropriation (Gleditsch et al. 2002). However, a linkage of environmental concerns with fights on the territory has been proved (Klare 2001), and competition and control over critical natural resources will be among the guiding principles behind the use of military force in the twenty-first century. The specific NTSIs [17] identified in the case of "environment-space" are several, depending from the perspective (military and/or civilians) and the referential object of security (humankind as a whole or single items put in space, being them satellites or crewed flights).

Some of them remained constant in time—as, for example, the threat of nuclear contamination from rocket launches—others, like debris, have become progressively worse and are listed among the most prominent aspects on non-traditional security.

There is specialised literature that looks at environmental NTSIs space. One of the most frequently cited is ESA (2015), which identified three sources of non-military threats to spaceflights:

1) space debris that can collide with satellites and spacecraft;
2) harmful "space weather", mainly generated by the Sun (solar flares, solar winds and so on) and, to a minor measure, cosmic rays;
3) potential strikes by natural objects, such as asteroids, that cross the Earth's path.

In this book, ESA's classification has been taken into account, with two integrations: one, the pressure of a growing presence in space—the scarcity mentioned above as one of the trigger in resource-related conflict; two, the issues of non-intentional contamination, meaning with that everything *not* coming out of an act of war[18] (such as the detonation of a device).

The world's space agencies have defined and set up plans to deal with each one of those NTSIs, as discussed below; debris, being such a huge and complex discussion, have been given a separate section.

A. Crowded Orbits: Scarcity in Outer Space

The concept of scarcity in outer space can be a tricky one to define. After all, is not space infinite by definition, or at least, perceived as such from our limited Earth-bound perception? Still, while space might well be infinite, usable Earth orbits to lodge satellites are not. What worse, these limited areas are running out.

There are three more commonly used Earth orbits, i.e. Lower Earth Orbit (LEO), the closest to the planet, Medium Earth Orbit (MEO) and geosynchronous orbit (GEO), all of them intensively used by satellites, depending on their uses and technical requirements. With a growing number of them, overcrowding has become an issue, and since a while; ITU has started warning states about the overpopulation of GEO since the beginning of the 1970s.

Now, the ITU Constitution explicitly states "Member States shall bear in mind that radio frequencies and any associated orbits, including the geostationary-satellite orbit, are limited natural resources and that they must be used rationally, efficiently and economically" (ITU 1992: Art. 44.2). In one recommendation issued in 1993 aimed at safeguarding GEO for satellite use, ITU[19] asked countries to discard satellites no longer in use into a sort of space graveyard, 235–450 kmabove GEO, in order not to fill up GEO with space junk. The situation in LEO does not look better, either, and the Inter-Agency Space Debris Coordination Committee has made LEO a designated debris-protected orbit, requiring spacecraft to abandon this protected space within 25 years of their end-life.

To give an idea of how quickly Earth's orbits can fill up, it is worth looking at the satellites sent into orbit in 2017 (complete statistics have been given in Chapter 3): of the 453 satellites sent into space, 403 went to LEO (including SSO), 40 to GEO and the rest to MEO (FAA 2018).

Table 6.2 Synthetic statistics of space object launched into space to date

Russia/USSR	3451
USA	3050
China	481
Japan	251
ESA	86
India	111
France	145
Germany	77
UK	102
Italia	40
Brazil	30
Total	8539

Source Author's elaboration on UNOOSA database, 2019

In order to better understand the dimension of the problem, it is also worth examining how many objects are out there. An UN-maintained registry of objects launched into space does exist since 1962, and "originally established as a mechanism to aid the United Nations Committee on the Peaceful Uses of Outer Space in its discussions on the political, legal and technical issues concerning outer space, the evolution of international space law resulted in space object registration becoming a means of identifying which States' bear international responsibility and liability for space objects" (UNOOSA 2018, online).

Those were the foundations from where the 1976 Registration Treaty (Chapter 7) and the requirement of registration have come from. As of today, over 91%[20] of all satellites, probes, landers, crewed spacecraft and space station flight elements launched into Earth orbit (or beyond that) have been reported in the UN Registry (Table 6.2).

Details of the satellite orbiting have been given in Chapter 3. Here is only the case to notice that, as of 30 November 2018, the Index reported 8303 objects, about 22% of these have been launched within the last eight years alone, and one out of four (2327) from US launching facilities. Russia accounted for more than one-third of them (3532), while China ranked third, with 452. Of all those objects, only 4919 of them were in orbit; these figures, however, do not include debris.

B. Harmful Space Weather

Many natural phenomena that go under the denomination "space weather". The most likely to occur is what is caused by solar activity. "The Sun emits energy, as flares of electromagnetic radiation (radio waves, infra-red, light, ultraviolet, X-rays), and as energetic electrically charged particles through coronal mass ejections (CME) and plasma streams. The particles travel outwards as the solar wind, carrying parts of the Sun's magnetic field with them" (Space Weather Canada 2018, online).

Radiation and particles reach the Earth's (geo)magnetic field with different moments and modalities, causing a series of effects, some of them innocuous (auroras), others harmful. In case of particular severity, CMEs are called "solar storms" and they are known to have debilitating effects on digital and technologic systems on the Earth's surface, and to satellites, disrupting radio communication and navigation devices.

Others, rarer types of space weather are related to cosmic rays. As the name suggests, they are a mix of high-energy photons and subatomic particles coming from outside the Solar System. The Earth is generally protected again them by the Sun itself, through the Sun's solar wind and its magnetic field. However, any time the solar activity varies, it exposes the Earth to potentially harmful quantities of cosmic rays. There is already evidence they can create problem to Earth infrastructure, such as aeroplanes.

To date, there has been only one instance they have damaged a Quantas flights, but many concerned studies have addressed this issue (Schwadron et al. 2014).

Space weather is closely monitored, and at least two organisations, the US Space Weather Prediction Centre of the National Oceanic and Atmospheric Administration[21] and the EU-based, ESA Space Weather Service Network,[22] keep track of the solar activities and offer hazard-prevention measures.

C. Asteroid and Threats from Outer Space

Asteroids[23] fully qualify for the Chinese word "weiji" [危机], crisis. An asteroid crisis and its prevention are at the basis of many ongoing initiatives, starting from the world's UN's International Asteroid Day on June 30, established in 2016 with the resolution A/RES/71/90 to raise awareness about the possible dangers of asteroid collision and NASA's plans and studies to deal with an emergency. There are a few projections about the likelihood of such events: "NASA's Near-Earth Object Observations Program says there have been no asteroids or comets observed that would impact Earth anytime in the foreseeable future. All known Potentially Hazardous Asteroids have less than a 0.01% chance of impacting Earth in the next 100 years" (NASA-JPL 2015, online).[24] Nonetheless, NASA has led a global effort to coordinate the detection of potentially hazardous asteroids (PHAs) and comets since 1998. NASA's ongoing Near-Earth Object Observations (NEOO)[25] Programme has already discovered over 90% of the near-Earth objects larger than one kilometre, and now it is focusing on finding 90% of the NEO population larger than 140 metres (Pravec et al. 2012). While unlikely, the event of an asteroid hitting planet Earth is not zero, and the date of Asteroid Day—30 June—has been chosen in remembrance of Tunguska Asteroid,[26] which hit Russia in 1908 and razed 2000 square kilometres of Siberian forest (with fortunately no human victims). It has been estimated that the blast was 185 times more powerful than Hiroshima.

As the two characters composing the Chinese word crisis WEI [危] = danger JI [机] = opportunity suggest, as much as any crisis per se contains both clear elements of dangers but also of renovation (and therefore opportunities) asteroids, now increasingly an object of studies, harbour both. Sections in the following chapters explore in detail space mining, in legal terms (Chapter 7), business concept (Chapter 8) and space exploration (Chapter 9).

D. Space Contamination

A final issue to be considered is the risk of contamination of space, always possible when rockets are used. An example was the (not assisted) re-entry of the Soviet satellite Cosmos-954, carrying an uranium-fuelled NPS on board. When it self-destroyed upon Canada, its radioactive debris threatened to contaminate cities and harm people. Albeit they eventually landed on a deserted area, the event was a sobering example of what can go wrong with space operations. Other examples, less likely but still possible, are the potential environmental hazards linked to organic contamination of outer space ("forward contamination"), which is going to be an issue with the forthcoming exploration of celestial bodies (Williamson 2010).

It exists also what the scientists call "back-contamination" of the Earth by unfamiliar contaminants transported by returning spacecraft, or by Earth-born, but space modified organisms. There are already studies on bacteria and other organisms on board of the ISS—there are plenty (Space.com, 10 April 2019a)—most of them found on Earth, but it is still too early if they represent any risk to astronauts. However, at least one of the microbes sequenced has shown an unknown DNA (Space.com, 2 January 2018).

More is to come.

6.5 NTSI: Debris in LEO and Space Readiness Plans

The name "space debris", also called "space junk" (IAA 2017; IADC 2007), is a wide category, and even technical definitions vary. And while their presence has long been widespread concern and many existing regulations have been enacted, those regulations were often limited in scope and not binding at an international level. However, even though country regulations neither work harmonically towards the same direction nor preview cooperation, they are in general consistent in their provisions. The most relevant, to date, is the United Nations Committee on the Peaceful Uses of Outer Space (UN COPUOS) and the EU-sponsored ESA initiative. A recent addition is the Inter-Agency Space Debris Coordination Committee (IADC) established in 1993 with the aim of strengthening research and coordination of member states according to the Guideline for Orbital Debris Mitigation in 2006.

The category of space debris includes objects that are no longer useful and not retrieved from orbit, parts of them included. Therefore, spent satellites, rocket bodies and fragments of them, solid material released during launches and any other parts floating out there increase their number. What does this junk come from? There are three main sources:

- breakups (accidental and intended);
- operational debris (often impossible to avoid);
- collisions, resulting in a break-up of the colliding parts.

While collisions still represent a minor part, they are bound to increase as the orbit become more and more crowded. This is especially true for LEO, which is where most of the activities take place and the smallest debris hover, raising the odds of collision (Wiedemann 2012). Debris production has not been uniform over the years, depending on a series of variables, such as the number of launches and the implementation of mitigation initiative. Over the 1990s until the early 2000s, the growth has been constant but with a linear function. Two events, one carried out intentionally, the other a pure hazard, changed it drastically:

- in January 2007, China conducted an anti-satellite (ASAT) weapons test and destroyed its polar-orbiting weather satellite Fengyun-1C[27];
- in February 2009, there was the first collision ever between two satellites, the (non-functional) Russian satellite Cosmos 2251 and the active US Iridium 33, an incident said to have "unchained a new chapter in the field of space debris" (Williams 2012: 333). The case has been studied in Chapter 7 for its legal implications.

The two events have produced about 5500 debris particles, which had been officially catalogued by the SSN. According to the estimates, 90% of which are still in orbit, and they account for 36% of all LEO objects. More worryingly, this kind of accidents can start the well-known, feared chain reaction of collisions called "Kessler effect" from the scientist that first identified it, which would make segments of the Earth's orbit no longer viable for human activities of any kind (Kessler and Cour-Palais 1978).

Detailed statistics are difficult, due to the growing number of debris in the first place but also to the different ways states catalogue them, based on their own working definition (Weeden 2011). For example, in 2011 the US Space Command's Space Surveillance Network (SSN), which classifies the ones of known origin, put this figure at about 16,000, with this breakdown: payloads: < 3500; fragments: 9000; rocket bodies: 1800; mission-related debris: 1800 (SSN 2011). To this number, however, they need to be added estimated 600,000 smaller (1–10 cm) uncatalogued objects in Earth's orbit and hundreds of million pieces within the range of one millimetre to one centimetre (US Congress 2012). It is worth noticing that the satellite and spacecraft shielding are only effective when debris is below 1 cm in diameter. Anything bigger than that represents a hazard, which gives the magnitude of the problem.[28]

If satellites are the first objects to be at risk, a crewed presence in space is threatened, too. The ISS had to manoeuvre more than a few times to avoid collisions, and it was possible only when the early warning system gave the astronauts enough time to execute avoidance manoeuvres. In the years between April 2011 and 2012, there were six instances of collision, four resulting in a manoeuvre; in two cases, the crew had to seek refuge in the emergency escape module (BBC News, 24 May 2012).[29] The frequency in the decade before was of about one per year.[30] This increase in avoidance manoeuvres is mirrored by the satellites, whose NASA and ESA have provided accurate statistics for the latest 15 years. The ESA initiative in particular is worth mentioning for the extent and the tools deployed to deal with this threat. ESA's Space Situational Awareness (SSA) programme has responded to these hazards by developing, together with a few European and international partners, a warning system that detects and forecasts these accidents. Moreover, as of December 2018, ESA has produced a debris map[31] monitoring newly found debris, which going to add into the Annual Space Environment Report. Only about 6–7% of tracked space objects are active, operational satellites. By the end of 2017, ESA estimates were of 19,894 bits of space junk circling Earth, with a combined mass of at least 8135 tonnes.

In addition to the tracking, there are a few mitigation initiatives[32] that have been put into place. One concerns the safe disposal of some of the most dangerous objects. As discussed in the ESA's latest report, the conditions of GEO are even worse than in LEO, due to the specificities of the geosynchronous orbit itself, which unlike in LEO has no natural sink mechanism by which objects could leave. Therefore, "the solar radiation pressure on the objects will also make long term predictions subject to non-negligible uncertainties. A payload or rocket body operating in the GEO Protected Region, with either a permanent or periodic presence, shall be manoeuvred in a controlled manner during the disposal phase to an orbit that lies entirely outside the GEO Protected Region" (ESA 2018: 63).

Procedures of clearance have been put in place, like "ESA's Clean Space Initiative", which looks at ways of cleaning up the space environment, as well as preventing new debris to be created in orbit. It is important to notice that even non-working space objects might constitute valuable assets (an inactive space object can be in reserve for future activities or carry valuable classified information; Perek 2005); therefore, their removal has either to be carried out by their public or private owner or, at the very least, previously agreed with them.

A way of dismissal is to lead end-of-life satellites into a "graveyard orbit", just above the GEO's altitude where, since 1970, a few of them have been steered into and put away. In addition to this not always feasible option, there is a "spacecraft cemetery" lying deep down in the southern Pacific Ocean, 3900 kilometres (2400 mi) southeast of Wellington, New Zealand. Some huge objects have been discarded in this way, like the Mir space station, and, more recently, Tiangong-1.

Together with ESA, Japan is one of the states that devoted attention the most to asteroids (and to the dangers of debris in LOE), committing for a decade a few probe missions for their study (e.g. Hayabusa missions. See Chapter 9).

A crucial aspect in the discipline of space debris is constituted by the legal liability aspects, which have become relevant in view of a recent case: the fall, in August 2018, of Chinese space station Tiangong to Earth. The event has been followed with attention both by the world press and space laboratories around the world, to calculate with some degrees of approximation where the debris would have followed and if they could cause any security hazard. As previously seen, the probability for injuries and damages in similar circumstances is low, but not statistically null. No harm resulted from Tiangong 1, but what would have happened if it did in terms of legal responsibility? The discipline has been examined in detail in Chapter 7, dealing with the legal aspect of the space sector, more specifically in Sect. 7.3.

Case Study: The ISS—The UN in Space

The ISS was not the first orbiting station, to be sure, and by no means the only one (it is one of the nine to date). But it is an example of a global world effort in space, combining the experience of the US Skylab (active between 1973–1979) and USSR/Russia Salyut (1971–1986) and Mir (1986–2000) (Fig. 6.1).

Fig. 6.1 The ISS

By far the longest servicing station (assembled starting from 1998 to become operational in 2000), it is also the most important in terms of the trove of science and experience that has delivered during its long years of services. The ISS represents, to date, the largest human-made body in Earth's orbit and the second most luminous object after the Moon; it can be easily spotted during its transits (cruising at an altitude in between 435 and 330 km, it completes 15.54 orbits per day) with the naked eye from Earth. The station has become over the years an important feature to the world's space perception. There are smartphone apps that track its course over the sky and teach people how to locate and identify it. Astronauts regularly tweet now from the ISS taking to the world space-community and sharing images and comments from up there.

Politically speaking, the ISS is considered the "most politically complex space exploration program ever undertaken" (NASA 1999), involving five national space agencies—NASA, Roscosmos, ESA, JAXA and CSA—while the usage protocols are regulated by international treaties. Since its launch, the ISS has carried out essential scientific research for the future of the humanity, and life, in space—such as experiments in microgravity, astrobiology, material sciences—together with helping research the field of astrophysics, astronomy and Earth sciences. There are several reasons for its success, the number of efforts made by the principal contributors to maintain good relationship being the first one, but also the scientific and financial commitment from each agency played a role.

In technical terms, the novelty of the ISS was its modular approach,[33] made as it is by different components from different space concepts, whose harmonisation constituted the first of a series of formidable challenges that the scientists needed to overcome to make the station work.

If 1998 marks the launch of the first module—the Russian Zarya—into orbit, the station had to wait until 2000 and the docking of the third component, the space habitat Russian Zvezda,[34] to be able to welcome its first human residents from Expedition 1 (ESA 2000). It has been, however, continuously inhabited since then. The amount of engineering work necessary to the assembly of the ISS can be appreciated if considering that, while with a few exceptions, all the Russian modules have docked robotically, the others had to be manually installed from the ISS using SSRMS (i.e. the station main robotic hand, the Canararm2) and EVAs (astronauts' extra-vehicular activities, aka "spacewalks").[35]

The ISS consists of two main sections, USOS and ROS, namely, the US and Russian Orbital Segments. The pressurised areas are accessible without a spacesuit, while the unpressurised superstructure of the station requires it. And if Zvezda provides computer guidance and navigation, the main lab and ISS research facility is constituted since 2011 by the US module Destiny (the other lab is the European Columbus).

The expansion over the years of the integrated truss and node structure—the skeleton that allows for units to be attached to the station—has made possible the addition of pressurised modules, such as the Japanese Kibo, another laboratory and the largest ISS module to date. As of December 2018, there are 15 pressurised modules, with five more, including the Russian Nauka, scheduled for launch in 2019.

A few spacecraft[36] serve and resupply the ISS. Since the decommissioning of the Space Shuttle in 2011, the astronauts have been taken back and forth by the Russian Soyuz, while cargoes have been shipped, among others, by SpaceX's Dragon. Differently from the previous stations, where simultaneous docking was limited or non-existent, the ISS allows for more than one ship to be in place—as the diagram below shows. On 10 January 2019, there were four spacecraft docked, the American Cygnus 10, the Russian Progress 70, 71 and Soyuz MS-11.

234 individuals from 18 countries have visited the ISS so far—the latest crew to be shipped is Expedition 58—with the USA (147 people) and Russia (47) providing the most numerous groups.

Accidents have occurred over the twenty years of ISS activities, even though less than it could have been expected considering how crowded LEO can possibly be. This fact also explains the number of corrective manoeuvres the station had to perform[37] over the years to avoid collisions or damage from debris. The latest instance was the recent accident in September 2018, when an unknown cause provoked a leak of air. An investigation about possible causes is still on the way and, while the leak was quickly identified and repaired, but there were a lot of rumours (back by a few Russians sources) and conspiracy theories of a sabotage from inside the ISS.

As of December 2018, there are only two space stations in Earth's orbit, the ISS and the Chinese Tiangong-2 (operational but not occupied on a permanent basis). However, the ISS days are numbered, albeit it has been extended due to the lack of viable alternatives. First, it was scheduled to be decommissioned in 2016, but after the extension, the retirement day is now set at some moment in the mid-2020s if no further extension is obtained.

Another interesting detail is that, a real marvel of combining different sections and space conception from different traditions, the dismantled ISS is going to revert to its original parts. Russia has already communicated the intention of retrieve sections of their original modules and put them together in a forthcoming experimental Russian space station. The new station will be called OPSEK—Orbital Piloted Assembly and Experiment Complex—and its main use will be to assemble spacecraft for human flight targeting the Moon and other planets in the Solar System. The rationale is evident: having an orbiting station in LEO will save tremendously in terms of fuel and costs.

Russia is not the only one with plans about future space stations. China, the most notable exclusion from the ISS, follows closely. After the grounding of Tiangong-1, sent in orbit in 2011 and back in 2018, the Asian giant has continued its investment in space stations. After Tiangong 2 (2016), Tiangong 3 will follow in 2022, with an estimated crew of three over 40 days. Other and more daring initiatives are on the way and, as it has been noted for crewed spaceflights, they come from the private sector. A new concept of space station has been proposed by Bigelow Aerospace, which, in cooperation with both NASA and SPACEX, aims at launching into orbit the first privately funded commercial space station—the B330—in 2020. In the company's plans, B330 will be able to accommodate a crew of six with more space—an internal volume of 330 cubic metres—than a typical semi-detached house. The project is already underway, and the prototype, Bigelow Expandable Activity Module (BEAM) is currently under testing—as it has been inflated and momentarily docked to the ISS. If successful, this new hybrid concept is going to make the realisation of orbiting space stations much cheaper than the current one and of a far easier setting up and replication procedures.

Sources

ESA. (2000, October 31). *First Crew Starts Living and Working on the International Space Station*. ESA 200. Available at https://www.esa.int/Our_Activities/Human_Spaceflight/International_Space_Station/First_crew_starts_living_and_working_on_the_International_Space_Station.

NASA. Available at https://www.nasa.gov/content/bigelow-expandable-activity-module.

Notes

1. Both factors have been analysed by the Copenhagen School when discussing non-traditional security issues (NTSI), whose importance is rapidly increasing in space matters.
2. If this enlargement of the traditional domain of space security is positive or even appropriate is a different issue, one that goes under the risks of securitisation. More about it in the conclusive section of this chapter.
3. It is impossible here, for reasons of available space, to compare two security approaches as difference as China and the USA. But any serious analysis of the future of security in outer space should start from it and from the importance differences they present. For more on the military dimension of outer space, see Dolman and Cooper (2011).
4. Susan Strange (1994) defines power through four important structures, and classes knowledge as ranking alongside production, wealth and security in terms of importance. It can be argued that space activity and exploration do occupy an important place at the centre of the IPE's structures of power, in terms of military, security, wealth, finance, production, knowledge, ideas and discourse.
5. In an enlightening discussion, Mutschler (2015) shows that all mainstream IR approaches come short in explaining this lack of cooperation in the traditional-intended space security. "Neorealism explains this lack of cooperation with the difficulties to achieve balanced gains; neo-institutionalism sees the establishment of effective rules and mechanisms to verify the compliance of states as a main hurdle; and from a Constructivist/Liberal perspective the main problem lies in the dominant beliefs about the value of unilateral space policies" (Mutschler 2015: 22). This is probably a telling sign that a new, and more comprehensive, security approach is needed in the space age.
6. In order to work properly in a preventive function, such SSA body should be placed other than on the Earth surface, probably in a NEO-orbit. This looks still far away in the future. As of today, apart from ESA and a few others, states lack SSA, and none is at the level of the US Space Surveillance Network.
7. There have been to ETS in project at those early times, one developed by the US Army (the Nike-Zeus system) and the other by the US Air Force (Project Thor). They became operational between the early 1960s and the mid-1970s, and both carried hydrogen warheads (Peebles 1983).

8. While WMDs are prohibited in space, the effects of nuclear explosions in the medium have been studied in 1960s—one of those projects was STARFISH—producing mixed results depending on a series of factors, but generally proving not optimal because of the indiscriminate damage (to domestic and enemy satellites) they provoked. This is why the research moved to a different kind of ASAT weapons, of the type used until today.
9. The term "missile" is also more open to interpretation that one can imagine in terms of characteristics of the device, considering its long history, about which the Encyclopaedia Britannica states: "There is no reliable early history of the 'invention" of rockets'. For more, see https://www.britannica.com/technology/rocket-and-missile-system.
10. For a more detailed analysis of those threats, see Pellegrino and Stang (2016).
11. This is why the new navigation systems pack anti-jamming equipment into them. See Galileo's case study in Chapter 4.
12. Tactical High Energy Laser (THEL). Not completely defunct, though. NG said that the technology has been used in the USA, for example, in the Army's Solid State Laser Testbed Experiment (White Sands Missile Range base, New Mexico). http://www.northropgrumman.com/Capabilities/ChemicalHighEnergyLaser/TacticalHighEnergyLaser/Pages/default.aspx.
13. According to the definition from the Encyclopaedia Britannica, a ICBM is a land-launched, nuclear head-armed missile with a range of more than 3500 miles (5600 km), first deployed by the USSR in 1958. See https://www.britannica.com/technology/ICBM.
14. "The absentee ratio is the number of weapons platform needed in orbit to have one available to engage a target within a desired timeframe. If a weapon system in LEO targeting Earth has a lateral range of 500 km on either side of the satellites ground track, and the weapon satellites are in polar orbits, there has to be 20 satellite orbits just to have the possibility to engage any target at the equator within 90 min. Since the de-orbiting time would be at least 10–15 min, the actual engagement time could be 105 min. If this is to be cut to about 30 min, the same time as the engagement time for an ICBM, the launch of the space weapon would have to happen within 15 min instead of 90 min. This can be achieved with six *satellites* in each of the 20 orbits, for a total of 120 satellites, or an absentee ratio of 120" (Hostbeck 2015: 976).
15. As China famously did in 2007, instantly creating a great portion of the debris currently in orbit. India did something similar in 2019, testing an

ASAT on one of its satellites (Space.com, 30 March 2019b). The legal case of debris has been discussed in Chapter 7.

16. It has been since the 1980s that the UN CD has considered the case of space weapons, but, as in other areas, no substantial progress has been made. Even when USA and China (having previously secured Russia's agreement) were ready to compromise on state security, Pakistan intervened and blocked the mandate for talks. (See https://www.un.org/disarmament/topics/outerspace.) Only a few areas have reached a consensus and a working agreement, such as in the case of debris.
17. It is worth to know that not a lot is explicitly devoted regulate activities from the point of view of the environmental issues mentioning in the non-traditional security issues mentioned in the previous chapter. There are a few provisions in the Outer Space 1967 and in the 1992 NPS Principles (Principles Relevant to the Use of Nuclear Power Sources in Outer Space), which are, to date, the most extensive under this aspect.
18. Everybody who remembers the disaster of Exxon Valdez and the oil spilt in the Arctic Sea understands how important this issue can possibly be. In the case of space, things might be even more crucial.
19. In 1996, the United Nations reported that severe crowding in the geostationary orbital slotsover Asia "led to the jamming of a communication satellite by PT Pasifik Satellite Nusantara (PSN) of Jakarta, Indonesia, in defence of an orbital position claimed by Indonesia. This incident focused global attention on the worsening problem of orbital crowding and caused the matter to be brought before the 1997 World Radio Communication Conference (WRC) of the 187 member-nation ITU in Geneva" (United Nations, Highlights in Space: Progress in Space Science, Technology, Applications, International Cooperation and Space Law, 1997) [Vienna: United Nations, 1998], 51.
20. As of 30 November 2018 (date of the latest statistical update), the Index contains data of satellites and other objects launched from 1957 to the present. However, space debris (the object of next section) and other non-functional objects are not included. Cu-Sat are not included either (only two are present, while the stats shown in Chapter 8 give a quite different picture).
21. See NOOA/SWE website: https://www.swpc.noaa.gov/ According to the records, the geomagnetic storm in 1989 was the most severed in the last 50 years.
22. ESA SWE website: http://swe.ssa.esa.int/.

23. For a definition of asteroids and the difference between natural satellites (such as The Moon) and comets, see, for instance, Space.com at https://www.space.com/51-asteroids-formation-discovery-and-exploration.html.
24. *NASA: There Is No Asteroid Threatening Earth.* (2018). *NASA/JPL.* 19 August 2015 Retrieved 7 December 2018, from https://www.jpl.nasa.gov/news/news.php?feature=4692.
25. NASA's Planetary Defence Coordination Office (PDCO) has facilitated communications between the astronomical community, the federal government and the public in the event that potentially hazardous objects are discovered.
26. Another, more recent accident occurred in dramatic proof that some of these may strike Earth came on 15 February 2013, when an unknown object travelling at 66,000 km/h and thought to be 17–20 m in diameter exploded high above Chelyabinsk, Russia, releasing 20–30 times the energy of the Hiroshima atomic bomb. The resulting shockwave caused widespread damage and injuries, making it the largest known natural object to have entered Earth's atmosphere since the 1908 Tunguska event, which destroyed a remote forest area of Siberia.
27. India's recent test in March 2019 has so far produced a smaller number of debris than the Chinese case, but it is too early to know for how long they will remain out there, or the threats they may represent to satellites. "Two days after the March 27 test, the U.S. Air Force said it was tracking 250 pieces of debris created when India's PDV-Mk II missile intercepted at 280 kilometres a satellite dubbed Microsat-R that India launched in January. On April 1, NASA Administrator Jim Bridenstine, lamenting an increased risk to the International Space Station, said that 400 pieces of debris were created" (SpaceNews 2019, online).
28. If this is difficult to figure out, it is because the environment is not taken into account. Impact velocities in Earth's orbits are enormous (0.1–0.8 km/s in geostationary orbit and up to 6–14 km/s in LEO). In this condition, even the tiniest particles become fatal, both to spacecraft and to astronauts involved in EVAs.
29. The incident has been reported here. ISS Crew Take to Escape Capsules in Space Junk Alert 2012, BBC News, 24 March 2012, www.bbc.co.uk/news/science-environment-17497766.
30. Increase in ISS Debris Avoidance Manoeuvers, 16–2 *Orbital Debris Quarterly News* (April 2012), 1, orbitaldebris.jsc.nasa.gov/newsletter/pdfs/ODQNv16i2.pdf.

31. There is a ESA Debris portal, which contains ESA's Space Debris Software. One is Debris Risk Assessment and Mitigation Analysis (DRAMA), a comprehensive tool for the analysis space missions. Another is Database and Information System Characterising Objects in Space (DISCOS). It "serves as a single-source reference for launch information, object registration details, launch vehicle descriptions, spacecraft information (e.g. size, mass, shape, mission objectives, owner), as well as orbital data histories for all trackable, unclassified objects which sum up to more than 40,000 objects" (ESA 2018, online). Both are available at https://sdup.esoc.esa.int/web/csdtf.
32. Executive office of the president, 14 August 2017. *Whitehouse.gov*. Retrieved 8 October 2018, from https://www.whitehouse.gov/wp-content/uploads/2017/12/08-14-17-OSTP-Orbital-Debris-Report.pdf. For more regarding the regulatory framework of debris, see Cox (2007). For mitigation initiatives, see Wiedemann (2012).
33. The Mir was a modular station, but only Russian.
34. Zvezda not only provided the first, essential life support for a crew of 6 astronauts. It also hosts (even today) the computer and the navigation system of the ISS, and, with its engines, provides boost for the station's orbit. (The alternative is to wait for visiting spacecraft and using their engines.)
35. This started with the second ISS component, NASA's Unity. The Unity node, which was to serve as a connector for other components of the station, was delivered by Space Shuttle Endeavour in 1998 during the first Shuttle mission to the ISS, STS-88. The Shuttle crew manually attached Unity to the Zarya during a EVA. By June 2011, astronauts had installed 159 components of the ISS during more than 1000 hours of EVAs (NASA 2011).
36. In addition to the ones mentioned here, there is also the Russian vehicle Progress and the Japanese H-II Transfer Vehicle.
37. See Celestrak for further details.

References

BBC News. (2012, May 24). *ISS Crew Take to Escape Capsules in Space Junk Alert 2012.* Available at www.bbc.co.uk/news/science-environment-17497766. Retrieved on 23 February 2017.

Bulkeley, R., & Spinardi, G. (1986). *Space Weapons: Deterrence or Delusion.* NYC: Polity Press.

Buzan, B., & Waever, O. (2003). *Regions and Powers: The Structure of International Security.* Cambridge, UK: Cambridge University Press.

Cox, L. V. (2007). *Avoiding Collisions in Space: Is It Time for an International Space Integration Center?* Carlisle Barracks: U.S. Army War College.

CSIS. (2018). *Space Threat Assessment 2018.* Available at https://aerospace.csis.org/wp-content/uploads/2018/04/Harrison_SpaceThreatAssessment_FULL_WEB.pdf. Retrieved on 21 February 2019.

Deudney, D., & Matthew, R. A. (1999). *Contested Grounds: Security and Conflict in the New Environmental Politics.* Albany: Albany State University Press.

Dolman, E. (2002). *Astropolitik: Classic Geopolitics in the Space Age.* London: Frank Cass.

Dolman, E. C, & Cooper, H. F. Jr. (2011). Space and National Security. Increasing the Military Uses of Space Increasing the Military Uses of Space. In C. D. Lutes & P. L. Hays (Eds.), *Toward a Theory of Spacepower.* Washington, DC: National Defense University Press.

Elliot, L. (2002). Regional Environmental Security: Pursuing a Non-Traditional Approach. In *Non-Traditional Security Issues.* Singapore: Institute of Defence and Strategic Studies.

ESA. (2015). Official Webpage. https://www.esa.int.

ESA. (2018). *Latest Report on Space Junk.* Available at https://www.esa.int/Our_Activities/Operations/Latest_report_on_space_junk. Retrieved on 18 February 2019.

European Commission. (2016). *Communication from the Commission to the European Parliament, the Council, the European Economic and Social Committee and the Committee of the Regions. Space Strategy for Europe* (COM 2016/705 Final), Brussels.

FAA. (2018, December 11). *The Annual Compendium of Commercial Space Transportation: 2018.* Available at https://www.faa.gov/about/office_org/headquarters_offices/ast/media/2018_ast_compendium.pdf.

Gallagher, N. (2015, July 30). The Logic for Space Arms Control. Bulletin of the Atomic Scientists Roundtable. https://www.cissm.umd.edu/people/nancygallagher. Retrieved on 2 January 2018.

Gleditsch, N. P., Wallensteen, P., Eriksson, M., Sollenberg, M., & Strand, H. (2002). Armed Conflict 1946–2001: A New Dataset. *Journal of Peace Research, 39*(5), 615–637.
Hardin, G. (1968). The Tragedy of the Commons. *Science, 162*(3859), 1243–1248.
Hebert, K. D. (2014). Regulation of Space Weapons: Ensuring Stability and Continued Use of Outer Space. *Astropolitics, 12*(1), 1–26. https://doi.org/10.1080/14777622.2014.890487.
Hertzfeld, H. R. & von der Dunk, F. (2005). Bringing Space Law into the Commercial World: Property Rights Without Sovereignty. *Space, Cyber, and Telecommunications Law Program Faculty Publications.* Available at http://digitalcommons.unl.edu/spacelaw/15. Retrieved on 10 January 2017.
Hostbeck, L. (2015). Space Weapons' Concepts and Their International Security Implications. In K.-U. Schrogl et al. (Eds.), *Handbook of Space Security.* New York: Springer Media.
IAA. (2017). *Space Debris Situation Report 2016.* Available at http://www.iaaweb.org/iaa/Scientific%20Activity/sg514finalreport.pdf. Retrieved on 16 December 2018.
IADC. (2007). *Inter-Agency Space Debris Coordination Committee Space Debris Mitigation Guidelines Update.* Available at https://www.iadc-online.org/Documents/UN_Presentation_2007.pdf. Retrieved on 16 December 2018.
INTA. (2018). *INTA Home Page* [Online]. Available at http://www.inta.es/opencms/export/sites/default/INTA/es/. Retrieved on 11 February 2018.
ITU. (1992). *Optional Protocol on the Compulsory Settlement of Disputes Relating to the Constitution of the International Telecommunication Union, to the Convention of the International Telecommunication Union and to the Administrative Regulations.* Available from http://www.itu.int/net/about/basic-texts/optional-protocol.aspx.
Jakhu, R. N., & Pelton, J. N. (Eds.). (2017). *Global Space Governance: An International Study.* Cham: Springer International.
Kessler, D. J., & Cour-Palais, B. G. (1978). Collision Frequency of Artificial Satellites: The Creation of a Debris Belt. *Journal of Geophysical Research, 83*(6), 2637–2646.
Klare, M. (2001, May 1). *The New Geography of Conflict.* Foreign Affairs. Available at http://www.foreignaffairs.com/articles/57030/michael-t-klare/the-new-geography-of-conflict. Retrieved on 18 December 2018.
Lawton, T., Rosenau, J., & Verdun, A. (2000). *Strange Power: Shaping the Parameters of International Relations and International Political Economy.* Aldershot: Ashgate.

Leissle, M., Hoerber, T., & Siglas, E. (Eds.). (2017). *Power, Politics and the Formation of International Law: a Historical Comparison, Theorising Space*. London: Lexington.
Lieberman, S. (2017). *Space Policy.* http://dx.doi.org/10.1016/j.spacepol.2017.04.004.
Lieberman, S., Hoerber, T., & Siglas, E. (Eds.). (2017). *Strange Spaces: An International Political Economy Reading of Operational Space Programmes, Theorising Space*. London: Lexington.
Mahan, A. T. (1890). *The Influence of Sea Power Over History.* Available at https://archive.org/details/seanpowerinf00maha. Retrieved on 10 February 2019.
Moltz, J. C. (2012). *Asia's Space Race: National Motivations, Regional Rivalries, and International Risk.* New York: Columbia University Press.
Moltz, J. C. (2014). *Crowded Orbits: Conflict and Cooperation in Space.* New York: Columbia University Press.
Mutschler, Max. (2015). *Arms Control in Space: Exploring Conditions for Preventive Arms Control.* London: Macmillan.
NASA. (2011). Official Agency Website, ISS Section Data.
NASA. (2015, August 19). *NASA: There Is No Asteroid Threatening Earth.* (2018). *NASA/JPL.* Available at https://www.jpl.nasa.gov/news/news.php?feature=4692. Retrieved on 7 December 2018.
NG. (2018). Official Company Website. http://www.northropgrumman.com/Capabilities/ChemicalHighEnergyLaser/TacticalHighEnergyLaser/Pages/default.aspx.
Nye, J. (1990). *Bound to Lead: The Changing Nature of American Power.* London: Basic Books.
Paladini, S. (2015). Resource Controversies Between Conflict and Cooperation: Agent-Based Models for Non-Traditional Security. In M. Neumann & D. Secchi (Eds.), *Agent-Based Simulation of Organizational Behavior.* Cham: Springer.
Peebles, C. (1983). *Battle for Space.* New York: Bookthrift Co.
Pellegrino, M., & Stang, G. (2016, July). *Space Security for Europe.* Paris: EUISS. Available at https://www.iss.europa.eu/sites/default/files/EUISSFiles/Report_29_0.pdf. Retrieved on 6 September 2018.
Perek, L. (2005). Management Issues Concerning Space Debris. In *Proceedings of the Fourth European Conference on Space Debris* (ESA SP-587) (p. 588). ESA/ESOC.
Pfalzgraff, R. (2015). International Relations Theory and Space Power. In C. D. Lutes & P. L. Hayes (Eds.), *Toward a Theory of Space Power.* London: Institute for National Strategic Studies.

Pravec, P., Harris, A. W., Kusnirak, P., Galad, A., & Hornoch, K. (2012, September–October). Absolute Magnitudes of Asteroids and a Revision of Asteroid Albedo Estimates From WISE Thermal Observations. *Icarus, 221*(1), 365–387. ISSN 0019-1035. http://dx.doi.org/10.1016/j.icarus.2012.07.026.

Sheehan, M. (2015). Defining Space Security. In Kai-Uwe Schrogl et al. (Eds.), *Handbook of Space Security*. New York: Springer-Verlag.

Space.com. (2018, January 2). *Astronauts Identify Mystery Microbes in Space for the 1st Time* [Online]. Available at https://www.space.com/39254-astronauts-sequence-microbes-dna-in-space.html. Retrieved on 10 April 2019.

Space.com. (2019a, April 9). *There Are Bacteria and Fungi All Over the Space Station, and Now We Know What They Are* [Online]. Available at https://www.space.com/microbes-fungi-space-station-identified.html?utm_source=sdc-newsletter&utm_medium=email&utm_campaign=20190415-sdc&fbclid=IwAR2Pn1YK3vSo-iQ4HoA1hBwmx4cdyBZXnPs7MDyXFK566R-LCNikRVoBVn4. Retrieved on 10 April 2019.

Space.com. (2019b, March 30). *India's Anti-satellite Missile Test Is a Big Deal. Here's Why* [Online]. Available at https://www.space.com/india-anti-satellite-test-significance.html. Retrieved on 11 April 2019.

Strange, S. (1994). *States and Markets*. London: Continuum.

Space News. (2019, April 9). *India ASAT Debris Spotted Above 2,200 Kilometers, Will Remain a Year or More in Orbit*. SpaceNews.com [Online]. Available at https://spacenews.com/india-asat-debris-spotted-above-2200-kilometers-will-last-a-year-or-more/. Retrieved on 10 April 2019.

Space Weather Canada. (2018). Official Webpage. Available at http://www.spaceweather.gc.ca/index-en.php. Retrieved on 16 December 2018.

Schrogl, K. U. (Ed.). (2015). *Handbook of Space Security*. London: Sage.

Schwadron, N. A., et al. (2014). Does the Worsening Galactic Cosmic Radiation Environment Observed by CRaTER Preclude Future Manned Deep Space Exploration? *Space Weather, 12*, 622–632.

Sempa, T., & Francis, P. (2018). The Geopolitical Vision of Alfred Thayer Mahan. *The Diplomat*. Available at https://thediplomat.com/2014/12/the-geopolitical-vision-of-alfred-thayer-mahan/. Retrieved on 6 December 2018.

SSN. (2011). Official Service Website. https://nssdc.gsfc.nasa.gov/nmc/spacecraft/display.action?id=1995-033B.

Sumida, J. (2015). Old Thoughts New Problems: Mahan and the Consideration of Space Power. In C. D. Lutes & P. L. Hayes (Eds.), *Toward a Theory of Space Power*. London: Institute for National Strategic Studies.

US Congress. (2012, January 27). *Active Debris Removal—An Essential Mechanism for Ensuring the Safety and Sustainability of Outer Space*. Report of the International Interdisciplinary Congress on Space Debris Remediation and On-Orbit Satellite Servicing.

UNOOSA. (2018). Official website. Available at http://www.unoosa.org/. Retrieved on 6 December 2018.

Voanews. (2018, October 4). *US to Confront Russia, China on Militarization of Outer Space*. Available at https://www.voanews.com/a/us-to-confront-russia-china-on-militarization-of-outer-space/4599793.html. Retrieved on 6 December 2018.

Weeden, B. C. (2011). Overview of the Legal and Policy Challenges of Orbital Debris Removal. *Space Policy, 27*, 41–60.

Weeden, B. C., & Chow, T. (2012). Taking a Common-pool Resources Approach to Space Sustainability: A Framework and Potential Policies. *Space Policy, 28*, 167–180.

Wiedemann, C. (2012). Space Debris Mitigation. In I. Marboe (Ed.), *Soft Law in Outer Space* (pp. 315–324). Cologne: Böhlau.

Williams, M. (2012). Space Debris as a 'Single Item for Discussion'. In *Proceedings of the International Institute of Space Law* (2011) (pp. 333–350).

Williamson, M. (2006). *Space: The Fragile Frontier*. Reston, VA: AIAA.

Williamson, M. (2010). Scope and Methodology. In M. Hofmann, P. Rettberg, & M. Williamson (Eds.), *IAA Cosmic Study 'Protecting the Environment of Celestial Bodies'*.

7
The Legal Dimensions of Space

7.1 The Fundamental Treaties of International Space Law

The existing framework for what concerns the domain of space law has its roots in the Cold War period and the negotiations between the two then-competing superpowers: the USSR and the USA.

The majority of the fundamental treaties that rule near and outer space have been negotiated at that time, often under the auspices of the UN. This is probably about to change, now that more and more countries have joined the "space club", and so did non-public actors.

However, if the past is of any guidance, the difficulties that have plagued the UN CD (Conference of Disarmament, which has been discussed later on in this chapter) represent a clear warning: every time countries seemed to agree on some important points, they were blocked by some of the (newer) players in the field. Therefore, any attempt to modify in one sense or another the few existing agreements, while representing an understandable evolution, might prove to be not necessarily sensible or even advisable, as this chapter argues.

S. Paladini, *The New Frontiers of Space*,
https://doi.org/10.1007/978-3-030-19941-8_7

The chapter opens with a general overview of the existing space legislation at an international level. A few of them were modelled on maritime and air regulations; not all of them are still in use; not all among the ones in today use have the same relevance[1] in terms of practical implications.

When discussing international space law, the most important transnational body is United Nations Committee on the Peaceful Uses of Outer Space (COPUOS), an international organisation established in 1959 (just after the start of the space era) with the UN resolution 1472 (XIV,[2]) "International Co-operation in the Peaceful Uses of Outer Space".

One of the aims of the Committee was "to study the nature of legal problems which may arise from the exploration of outer space", and, a few years later, this small expert unit moved under the Department of Political and Security Council Affairs. In 1992, the Division became the Office for Outer Space Affairs (UNOOSA) with HQ in the UN Offices in Vienna, of which COPUOS is now part (Fig. 7.1).

During its almost sixty years of activities, the Committee concluded five international treaties and established five sets of principles on space-related activities. As stated by the UNOOSA, "each of the treaties stresses the notion that outer space, the activities carried out in outer space and whatever benefits might be accrued from outer space should be devoted to enhancing the well-being of all countries and humankind, with an emphasis on promoting international cooperation" (UNOOSA 2018, online).

What is generally referred to as "space law" is actually made of five treaties,[3] whose application the UNOOSA oversees. Due to their relevance for the topics covered in this book, these treaties have been discussed in detail in Sects. 7.2 and 7.3.

I. The "Outer Space Treaty" (OST)

Treaty on Principles Governing the Activities of States in the Exploration and Use of Outer Space, including the Moon and Other Celestial Bodies.

Adopted by the General Assembly in its resolution 2222 (XXI), opened for signature on 27 January 1967, entered into force on 10 October 1967.

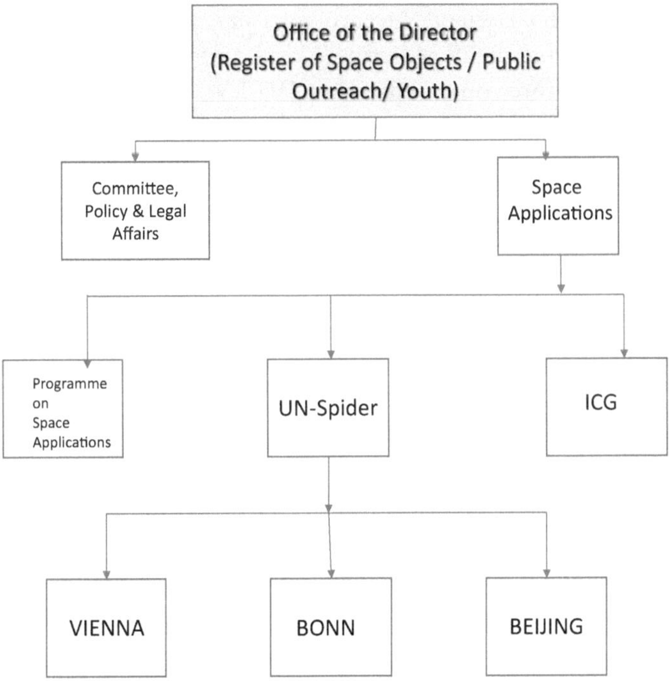

Fig. 7.1 Structure of UNOOSA

II. The "Rescue Agreement" (RA)

Agreement on the Rescue of Astronauts, the Return of Astronauts and the Return of Objects Launched into Outer Space.

Adopted by the General Assembly in its resolution 2345 (XXII), opened for signature on 22 April 1968, entered into force on 3 December 1968.

III. The "Liability Convention" (LC)

Convention on International Liability for Damage Caused by Space Objects.

Adopted by the General Assembly in its resolution 2777 (XXVI), opened for signature on 29 March 1972 in London/Moscow/Washington and entered into force on 1 September 1972.

IV. The "Registration Convention" (RC)

Convention on Registration of Objects Launched into Outer Space.

Adopted by the General Assembly in its resolution 3235 (XXIX), opened for signature on 14 January 1975, entered into force on 15 September 1976.

V. The "Moon Agreement" (Moon)

Agreement Governing the Activities of States on the Moon and Other Celestial Bodies.

Adopted by the General Assembly in its resolution 34/68, opened for signature on 18 December 1979, entered into force on 11 July 1984.

As any international treaty is as good as it is widely recognised and applied by the signatory states, it is important to keep in mind that not all the five treaties have been equally ratified.[4] OST is the one ratified by the majority of countries with space interests (see next section); more importantly, it also provides all the others (namely, the rescue agreement, the liability convention and the registration convention) with the legal basis for their own provisions.

The lack of important signatures has at times hindered the importance and the applications of the principles included in international treaties. That was the case of the 1979 Moon Treaty, generally considered a failed agreement since all the important players in outer space have not signed—the USA, China, Russia, Japan, UK—or ratified it—France, India. As a result, this agreement has obtained the most limited consensus.[5]

There are other agreements, which need to be mentioned here for the importance they had in terms of security, as briefly discussed in the previous chapter. The following ones have dealt in a way or another with disarmament and denuclearisation of outer space.

1. 1963 Partial Test Ban Treaty[6]

This was the first convention adopted by the UN with the Resolution 1884, prohibiting since that early stage nuclear explosions, *inter alia*, in outer space. "*While not banning tests underground, the Treaty does prohibit nuclear explosions in this environment if they cause "radioactive debris to be present outside the territorial limits of the State under whose jurisdiction or control" the explosions were conducted*" (US State Department, 2018, online)

2. 1972 Anti-Ballistic Missile Treaty

This treaty was negotiated between the USA and the Soviet Union as part of the Strategic Arms Limitation Talks, and addressed the issue of ICBM. Since the USA withdrew in 2001, the treaty is defunct.

3. 1981 PAROS—Prevention of Arms in Outer Space

This treaty was signed to address a legal loophole in the previous treaties dealing with outer space. While those agreements forbid weapons of mass destruction in space, they were previewing nothing about the use of other types of arms in space, particularly ASATs and other new kinds of sophisticated weapons. 1981 PAROS treaty was the answer to it.

In 1985, the UN Conference on Disarmament (CD) formed an *ad hoc* committee to address issues regarding PAROS[7] "*in order to prevent an arms race in outer space*".

In addition to the above-mentioned agreements, there are other legal documents[8], declarations and regulations that, even when not being specifically addressed to the space sector, do contain important provisions.

UN Resolutions over space activities and related sectors constitute a large part of them. Others are the acts coming from other parts of the UN system and other international bodies.

One of the most important is the International Telecommunications Union (ITU), which is the body responsible for allocating radio frequencies and orbital locations to the increasing number of satellites that provide public and commercial services. As for the UN General Assembly Resolution No. 1721 (XVI) of 1961, which requested ITU to report on those aspects of space communications requiring international cooperation, the international body produced its first *Report on Telecommunication and the Peaceful Uses of Outer Space* (1962), and after that, every year.

Finally, the five principles.

A. The "Declaration of Legal Principles"

Declaration of Legal Principles Governing the Activities of States in the Exploration and Uses of Outer Space.

General Assembly Resolution 1962 (XVIII) of 13 December 1963.

B. The "Broadcasting Principles"

The Principles Governing the Use by States of Artificial Earth Satellites for International Direct Television Broadcasting.

General Assembly Resolution 37/92 of 10 December 1982.

C. The "Remote Sensing Principles"

The Principles Relating to Remote Sensing of the Earth from Outer Space.

General Assembly Resolution 41/65 of 3 December 1986.

D. The "Nuclear Power Sources" Principles

The Principles Relevant to the Use of Nuclear Power Sources in Outer Space.

General Assembly Resolution 47/68 of 14 December 1992.

E. The "Benefits Declaration"

The Declaration on International Cooperation in the Exploration and Use of Outer Space for the Benefit and in the Interest of All States, Taking into Particular Account the Needs of Developing Countries.

General Assembly Resolution 51/122 of 13 December 1996.

A few recent initiatives at an international level have been examined when talking about the international framework. The rest of this chapter deals instead with the more general legal framework of space activities, intended for non-military (i.e. civilian) purposes.

"The year 1967 represented a milestone in space history with the entry into force of the foundational instrument of international space law: the Treaty on Principles Governing the Activities of States in the Exploration and Use of Outer Space, including the Moon and Other Celestial Bodies" (UN 2017: 1).

This statement by Simonetta Di Pippo, present Director of United Nations Office for Outer Space Affairs, is not an overstatement. In terms of relevance and importance, none of the treaties overtakes the first one, 1967 OST as it is generally referred to, and that is why it is necessary to examine its provisions first.

7.2 The "Outer Space Treaty" (OST) of 1967

UN Outer Space Treaty (OST) states the cardinal principles and directives for human activities in space and provides all basic framework for them. Originally signed by the Russian Federation, the UK and the USA in January 1967, it entered into force in October 1967, and it is the one ratified by the largest number of states. As such, it is not unfair to say it is the most important piece of legislation as far as space law goes. "*Indeed, the treaty has (thus far) successfully created an environment that has prevented warfare in space. Its binding provisions are not only legally defensible, but have also historically reinforced an overwhelming political dynamic to refrain from overt military action in space*" (Freeland and Blake 2017, online). Freeland also explained how the regulatory framework of space law—OST and all the following treaties—extended the principles of international law (*jus in bello*) to space activities as well.

As of October 2018, 107 countries have ratified OST, while 23 have signed the treaty, but the ratification is still ongoing. Most importantly, all the countries significantly involved in space activities did.

The treaty represents as such the main source of space law, which rests on the following principles:

1. Freedom of use, exploration and scientific investigation of outer space open to everybody (Art. 1)[9] and has to be carried out for the benefit of all states. Hence, the definition of outer space as "*province of all mankind*".
   This seemingly plain and generic provision has nonetheless created issues when it comes to the width of its application: Should all the activities be carried out for the benefit of all the states? If this was the case, any military activity that takes place in outer space, especially ASAT launches (see Chapter 6), should therefore be considered illegal as it violated this article. That is, unless the phrase is to be intended in a more generic way (and on the interest of the international community as a whole; Chatterjee 2014).
2. No state can claim sovereignty of outer space, by use, occupation, or other means (Art. 2).[10] This important principle defines outer space as res communis and, as such, it can't be object of appropriation. However, this provision can become problematic when it comes to commercial activities, such as space mining. This will be discussed in one of the next sections, devoted to national laws.
3. All states must act towards maintaining international peace in outer space, which essentially means "not to place nuclear weapons or other weapons of mass destruction in orbit or on celestial bodies" (Art. 3), and using celestial bodies for "exclusively for peaceful purposes" (Art. 4). This is probably the "single most important provision in the space treaty system that deals with demilitarisation of outer space, thus posing a potential legal challenge to the deployment of ASATs" (Chatterjee 2014: 29). However, the letter of the treaty is ambiguous, and when talking about the prohibition of the establishment of military outposts, the treaty is silent above the Moon (while explicitly mentioning the Moon in other parts of the treaty. This has raised doubts about the inclusion of the Earth's natural satellite into it; Gorove 1973). In any case, there is at the moment a strong international pressure against the weaponisation of outer space, as mentioned in Chapter 6, and restrictive measures in this sense are under discussion.

4. All states must regard astronauts "as the envoys of mankind", and therefore give assistance to them in any "event of accident, distress, or emergency landing on the territory of another State Party or on the high seas" (Art. 5).
5. States are responsible for all their national activities in outer space, no matter if it is not the government that directly carries them out (Art. 6).[11]
   This is another problematic provision, coming from the early days where only states were active actors in outer space, but that today is bound to create endless issues. This is also the area where the national space laws are now filling with directives and regulations in order to clarify and provide a framework for private and commercial activities.
6. States are directly liable for damages caused by their space objects. More precisely, the OST identifies two sources of responsibility in terms of damage liability: (a) the States that "launches or procures the launching of an object into outer space" and (b) "the State whose territory or facility an object is launched" (Art. 7). As seen in Chapter 6, the two cases do not necessarily point to the same organisation/country, which complicates the application of the norm.
7. The ownership of an object launched into space depends on whose registry it resides (Art. 8),[12] which it is not necessarily the launching state (even though at the time of the treaty that was the case). Moreover, no matter where those objects land when getting back to Earth, they belong to the original owner, to which they have to be returned, even in "component parts".
   This is another example of an apparently uncontroversial measure that is now raising many issues. Chapter 6 has discussed at length space debris and all the problems their handling do create. Now, since any debris is, as the spacecraft it comes from, a property of their original owners, they can theoretically only be removed or destroyed by them. A possible solution would be to allow, under the Rescue Agreement, the removal of space debris, no matter their "nationality", by a threatened stakeholder (e.g. a satellite or a spacecraft risking collisions), without this constituting a sovereign breach. However, OST does not preview anything in kind, and there have been no provisions in this sense.

8. States shall avoid harmful contamination of space, celestial bodies and planet Earth as a result of their space activities (Art. 9[13]).
   A lot has been written regarding this provision, for many reasons. The most evident is in reference to ASATs and in general weapons in Earth orbit, as discussed before. But it is not the only domain of application. Any exploration of another celestial body opens the way to two kinds of contamination, one by extraterrestrial organisms (or Earth-mutated organism returning from other space) and one by Earth's bacteria to previously pristine environments.
   An example of the first case is a study carried out at NASA's Jet Propulsion Laboratory (JPL) in California, where a team analysed bacterial samples collected from around the ISS in 2015. The analysis found "four previously unknown strains of antibiotic-resistant bacteria" (Singh et al. 2018, online). The second case represents a serious concern for space exploration; a series of protocols have been devised to address this danger, as, for instance, the measures taken for exploring Mars, adopting what NASA calls "planetary protection technologies".[14] This is one of the reasons why, when in 2015 NASA's rover Curiosity detected likely signs of water under Mars's surface, it did not investigate further. An exploration of something as delicate as a water source requires sterile probes, which NASA is currently working on (see Chapter 9). The article also recommends prior international consultations in any instance where space activities may deem to be harmful or causing "*harmful interference in [others'] space activities*". This specific provision can be ambiguously interpreted (Mineiro 2008), since it does not cover, for instance, all kinds of ASAT activities that are not conducted in outer space but still targeting objects in outer space, such as satellites in Earth's orbit. Moreover, since there is no accepted definition of "harmful interference[15]" in any space treaty, the domain of application remains vague.
9. States should cooperate in space activities (Arts. 10–11).
   Directly descending from the principle that outer space is a "province of all mankind" is the encouragement to cooperation in all the possible forms, as allowing other states to "observe the flight of space objects" launched by others (Art. 10), the dissemination of information about states' activities (Art. 11) and the opening of their own facilities—

"stations, installations, equipment and space vehicles"—to the visit of other State's representatives.

With all the limitations noted here, OST does represent the basic text of space law, and all the successive regulations, both national and international, made references to its principles.

7.3 Regulating Space Activities

After 1967 OST had stated the broad, general principles for the pacific use of outer space, it was soon evident that those principles needed to be integrated and developed into regulations that could be directly applied to space activities. The four OST "children" are the other four treaties contained in the UNOOSA canon, that addressed at one title or another, essential aspects of the space law.

The Rescue Agreement, signed in 1968, regulates the rescue and return of men and objects launched into space according to the principle already mentioned in OST, mirroring the existing regulations about international waters.

Signed in 1975 and entered into force in 1976, the Registration Convention[16] requires the notification of the launch of any space object into the Earth's orbit or beyond, and its inclusion in the UN-maintained registry of the space objects. Under these provisions, all the states are "*required to establish their own national registries and provide information on their space objects to the Secretary-General for inclusion in the United Nations Register. Responsibility for maintenance of the Register was delegated by the Secretary-General to the United Nations Office for Outer Space Affairs*" (UNOOSA 2018, online).

It has already mentioned that the 1979 Moon Treaty is generally considered a "failed treaty".

Some have argued this is a pity, because the Moon Treaty not only closed an important gap in 1967 OST (see further), but also represented a more comprehensive treaty than the previous one in terms of peaceful use of space bodies, defined, as in 1982 UNCLOS, "*Common Heritage of the Mankind*" (aka CHM[17]).

However, the treaty, with its CHM definition, raised some issues particularly difficult once both government and industry of various countries began exploring the possibility of the utilisation of natural resources from outer space. Listner has discussed the Moon Treaty is a series of articles, explaining how the treaty maintains that the lunar soil and its natural resources are the common heritage of mankind. Therefore, the harvesting of those resources has to be considered forbidden except when an international regime is established for the exploitation of such resources. In the Treaty itself, "*the exact nature of this regime is not detailed, nor is the term "resources" defined. It is reasonable to presume that the term "resources" would include recently discovered mineral deposits including titanium, the substantial water ice discovered at the Moon's south pole, and the helium-3 within the lunar regolith that entrepreneurs such as Apollo 17 astronaut Harrison Schmidt have proposed to extract to power future fusion reactors*" (Listner 2011, online). However, as it has been noted, even though the Moon Treaty has not been signed by the majority of the space-active countries, it is still valid international law in all its aspects, and it could be a game changer if one of those states decided eventually to ratify it.

One of the main issues with international space law is that the abovementioned treaty does not provide a state-to-state dispute settlement mechanism, with only two exceptions. The Liability Convention of 1972 is the only treaty that contains any mechanism whatsoever of international dispute[18] settlement in space, namely provisions for dealing with damages caused by space objects and definition of parties' liability. As mentioned before, this aspect is becoming increasingly important due to the growing problem of space debris and their handling by national and international bodies.

First of all, the LC treaty re-asserted the liability of states for damages caused by space objects already contained in Art. 7 of OST. The responsibility herein considered has neither value ceiling nor can be limited in time or territory. The rationale here is clear: the legislators wanted to force states to take all the possible measures to limit damages caused by their space activities. The LC went further than OST, distinguishing between two kinds of liability.

The first one is the so-called "absolute liability", covered by Art. 2: "*A launching State*[19] *shall be absolutely liable to pay compensation for damage caused by its space object on the surface of the Earth or to aircraft in flight*". Therefore, claims have not to demonstrate any actual fault of the launching state: the damage made is enough evidence.

However, the treaty previewed another concept of liability, the so-called "fault liability", which is intended for damages happened elsewhere, such as space (Art. 3).[20] In this case, the treaty limited liability: a state is liable "*only if the damage is due to its fault or the fault of persons for whom it is responsible*". Otherwise said, only if the fault is proved by the other part.

The LC went as far as previewing a procedure for compensation claims and dispute settlement. Significantly, LC article 8, which lay down the legal ground of the claim, restricted it to the nation states. "*A State which suffers damage, or whose natural or juridical persons suffer damage, may present to a launching State a claim for compensation for such damage*" (Art. 8). Therefore, private corporations, while still entitled to damages, have no independent rights to compensation and they cannot claim them by themselves. Acceptable as it was at the time the treaty was signed, it presents dramatic shortcomings nowadays that the activities of private space operators are on the rise. Moreover, not only privates were excluded from the access to the Claim Commission; international organisations and non-government organisations could not make appeal to the convention either. And while it can be argued that governments have all the interest in supporting their national champions, states are not legally bound to accept the request of submitting claims on the private companies' behalf.

Time and modalities of claiming represent an additional issue: since the recourse has to go through diplomatic channels (Art. 9), in a timely fashion (Art. 10[21]), the likelihood for privates to be able to successfully activate the treaty becomes even lower. The dispute settlement is straightforward. First of all, all the parts involved are invited to address their disagreements through negotiation, and if no agreement is reached, then either part can be a Claim Commission (Art. 14). The appointment of the Commission and any escalation of the issues were also stated (Art. 15).[22]

There are, however, a few limitations to this mechanism: first of all, the decisions of the Claims Commission preview no enforcement mechanism: they are recommendatory in nature, which of course limit their effectiveness. Another problem is that only what is included by LC definition of damage can be taken for a claim, and that definition is far from being all-inclusive.[23]

Before having a look at the national state laws in the next section, it is important to remember the LC stated that the discipline of liability could not be altered by future national space law, which may prove a hurdle for private-led space endeavours. However, states were allowed "*a right of recourse, if the state has paid compensation for damage caused by a space object operated by a non-governmental operator. Frequently, states demand that the operator takes out insurance in order to guarantee that the amount is also recoverable in practice*" (Marboe 2015).

The number and level of disputes are expected to increase steadily in the coming years, following the continuous growth of the economic value of space activities and the number of the public and private actors which take parts into them. Many observers, both in academia and in business, have warned that the absence of mechanisms of dispute settlement more robust than what previewed by LC might have the unintended consequence of weakening the applicability of space law as a whole and makes the whole regulatory framework more uncertain.

One of the solutions[24] proposed is the *Optional Rules for Arbitration of Disputes Relating to Outer Space Activities, adopted by the Permanent Court of Arbitration (PCA)*[25] on 6 December 2011 in Le Hague.

Those rules, based on 2010 UNCITRAL Arbitration Rules, are of particular importance because they reflect how activities in space have evolved since the years of LC, including in their provisions parties before excluded, such as international organisations and private actors. They also established a voluntary but binding regime, which leaves the decision to an arbitration body and previews procedures to appointing one.

Considering that, as of November 2018, the Permanent Court of Arbitration (PCA) was counting 121 among its contracting parties (PCA 2018),[26] those provisions seem of a certain interest. Moreover, by providing contractual formulas that can be inserted into specific contracts between parts, the PCA's Adoption has done a lot to provide legal certainty in a sector where ambiguity and the lack of clear, enforceable regulations are rampant.

In the word of Tronchetti, "*whether they will be able to accommodate the needs of space actors and prove to be effective will be known only once governmental and non-governmental entities start relying on them. Nevertheless, there are already signs that suggest that space actors will progressively increase reference to the Rules*" (Tronchetti 2009: 18), and, therefore, probably use them upon the occurrence of space-related disputes sooner than later.

7.4 International Trade and Space

As shown in the previous chapters, trade exchanges directly or indirectly connected with the space sector have soared, and they all fall under the remit of WTO's rules in the absence of more favourable provisions (WTO 2017). While international trade as a whole started to be regulated since the end of the Second World War by the (ever-evolving) institutional framework set up at Bretton Woods and modified by successive Rounds (Jackson 1969), the WTO itself has not discussed the specifics of outer space, or, more in general, the entire space business, until recently. There is a reason for that: for decades, the space industry has been limited in scope and players and covered by the same provisions of the defence sector. A commercial and civilian-oriented space industry, where private operators emerged together with the traditional, government-funded companies and space agencies, was essentially a novelty of the 1980s, and regulations so were regulations (McCormick and Mechanick 2013; Walter 2011).

Those changes occurred, however, not in the space industry as a separate sector but included in the more general sector of telecommunications, whose satellite business was part. Since then, there have been many discussions on whether the satellite business should be even included in the WTO framework. To date "none of those sectors, so far, has any such discussion given rise to actual and formal involvement of GATT, GATS and the WTO" (Dunk 2015: 818).

While the trade of actual goods—like equipment and their parts—is more straightforward, not so is the case of services; as a matter of fact, the application of GATS provisions or other multilateral regulations to the space sector has not happened as of today, with the only exception of satellite communication sector.

Apart from telecommunications, the other sector of relevance in terms of services in the space sector is the commercial launches, whose landscape is still fluid and in full evolution (Hansson and McGuire 1999).

Figures and outlook of the launching sector have been discussed in Chapters 2 and 3. The sector itself has seen, during and the 1980s and 1990s, a flurry of activities with the establishment of a global market for commercial launch services (Scarborough 1992; Scott 1996); and yet, it still looks in need of better, international applicable regulatory framework, since the legal provisions vary a great deal from one country to another, hindering exchanges.

In Europe, for example, the whole space sector, including satellite communications, is far more integrated, since the EU pushed the implementation of pro-free trade legislation in a way unheard elsewhere. On the other hand, this regional liberalisation has not extended across blocs. Both the USA and the EU, mirroring a policy already existing in the aircraft manufacturing sector, do restrict, even though in a different way, their own internal satellite market to non-national companies (Lai 1993; Horsley 2005). In the case of the USA, it is full-fledged closure; in the case of the EU, countries belonging to ESA prefer favouring their own providers (as with the Ariane rocket) rather than forbidding the use of other launchers by legally binding provisions.

The emersion of more players in the launching sector has, of course, added another dimension of competition. In 1980, on the wave of the 1979 opening to a market economy promoted by Deng Xiaoping's reforms (with the establishment of SEZs in Guangdong and Fujian provinces), the Chinese government created the SOE Great Wall Industrial Corporation (GWIC) and began offering launching services to foreign satellite operators in 1985 (He 1991; Pike 1991).

In 1992, the fall of Soviet Union opened up the space market through the new-established CSIs, with operators in Russia and Ukraine exporting the state of the art of Russian space technology at a considerably lower price (but quality consistently higher) than its Chinese counterparts, raising concerns in the Western countries (Twigg 1994; Zaborskiy 2006).

As a result, in the 1990s the USA adopted a case-to-case, bilateral approach (Burnett and Schroeder 1994; Burnett and Lihani 1997) to deal with the proliferation of private operators, instead of working on a multilateral regime along the lines of what GATS had started to promote since 1994. Moreover, even though many of the bilateral agreements signed in that period were negotiated taking into consideration the GATS rules, they don't reflect GATS regulations. This was due partly to military concerns and partly to protectionist fears; maintaining a healthy demand for their domestic launching services constituted a priority for the US government. There was, as such, a common denominator to all these agreements, and it was in terms of restrictions unilaterally imposed by the USA in terms of the quota of international commercial satellite launches (including satellites with US components) companies from those states were allowed to bid for, and a threshold—a minimum price they were obliged to ask—which had somehow to reflect Western provided.

The overall result has been the creation of an embryonic regime allowing for quotas and threshold prices but still somehow providing a global framework for the commercial launching services around the world. Interestingly, Europe kept outside this bilateral negotiation agreements and, to date, the EU has still not signed any in this sense, albeit Arianespace has been involved in a few disputes[27] since the onset of its activities.

There are reasons for optimism, though; this peculiar regime is due to change in time, as more and more operators enter the market and the market itself changes character and scope. The USA has allegedly stated the intention to enlarge the system bilateral agreements to build a global framework, including the EU, the most notable absence to date, even though the fact that none of the agreements has been renewed in the 2000s might suggest otherwise.

There is international pressure to include a specific annexe in the GATS services for the launching services, albeit nothing happened to date yet. It does not mean, however, that it is not going to happen in the future; in a scenario with multiple players private and public an international framework regulated at WTO level might prove the simplest of the solutions. It is to be expected in the coming future that GATS provisions will apply to a few segments of the space industry, most likely commercial and civilian-led activities.

Dunk argues that "*if economic size and international economic interdependency (of markets, consumers and producers, and traders) are the key criteria for becoming eligible for inclusion in the WTO's international trade frame-*

work regime, there would presently be at least four sectors of space activities and closely related space applications where such a case could be made (with a fifth possibly around the corner): launching, satellite communications, satellite remote sensing and satellite navigation, plus private commercial spaceflight" (Dunk 2015: 818).

This shift from national regulations to a GATS-level framework can happen sooner than later, due, for instance, to the growing presence of small satellite developers. Due to the scarcity of launching slots—the demand and supply in the peculiar launching market have been analysed in Chapter 3—there are plenty of developers in the US market that are obliged to piggyback on other launches, instead of being the primary payload[28] (which grants, as shown before, a series of privileges).

While in the past there were no viable alternatives, there are now a few international launchers—such as India's Polar Satellite Launch Vehicle (PSLV)—that can cater to this demand. Just in February 2017, the PSLV launched 104 foreign satellites into orbit (WSJ, 15 February 2017), overtaking the Russians for the biggest number of satellites sent out at one time. As a result, many US developers are now lobbying to lift the ban on the use of foreign launchers for US companies (De Selding 2016). The US approach so far has been to offer India the same conditions of other bilateral agreements previously signed—the already mentioned minimum price—but India refused (Listner 2016).

An international legal framework based on GATS[29] rules would successfully address, and probably solve, this conundrum, and others in kind that plague a sector still full of restriction. As it stands, India is going to benefit because the USA mandates that India's prices have not to fall below a certain minimum is a quota. However, "*quotas are prohibited in the GATS (so long as the United States does not sign a limitation into its schedule of commitments, an act that would defeat the whole purpose of negotiating). The United States stands to benefit for a similar reason: India's subsidisation of its PSLV is counter to the letter and spirit of the GATS*" (MJIL 2017, online).

Another important benefit of adopting GATS in the space would be its internal dispute settlement mechanism (thanks to the WTO), which will quell any worries that one party might take unfair advantage of the agreement, and set a common ground where commercial regulations and international WTO-related provisions would fully apply. This is not a minor point, considering what already mentioned regarding the existing dispute settlement provision existing in the space sector,[30] due to its non-binding character and the restricted, or non-existent, access to these settlement procedures to private companies.

The few cases until today prove the formidable challenge.

At the moment, the segment of the space sector that seems the most active in trying to promote a common framework is the one looking at exploiting space resources (see Chapter 8). In terms of international efforts, it has led to joint initiatives, such as the sessions of the Legal Subcommittee[31] of UNCOPUOS or to the informal meetings of The Hague Space Resources Governance Working Group[32] and their 2017 framework draft for the development of an international framework on space resource activities.

7.5 The States and Their Citizens: National Space Laws

The need of national space laws is becoming more evident by the day.

"*While the fundamental principles set out in the treaty are vitally important to the peaceful and orderly use of outer space, the pace of development of space-related technology – which allows for activities far beyond the contemplation of those that put the treaty together – means that some activities in space may fall between the cracks. [...] As space has become more accessible and commercialised, those broad statements of principle are, in our view, still necessary but no longer sufficient. They need to be supplemented*" (Freeland and Blake 2017, online).

This is not an isolated opinion: many believe that the 50 years old OST needs to be updated, and rapidly. A way would be ratifying the Moon Treaty or signing another comprehensive agreement that represents the changed outlook of the sector. In the meantime, countries are equipping themselves to deal with the growing presence of private space operators and with the new opportunities arising.

National space laws are a recent thing: nothing was needed when the space business was a matter of government initiatives and budgets. But it does not mean those provision cannot be beneficial also for state-owned enterprises and space agencies as well, due to the different characteristics of national laws and international laws. As a matter of fact, national laws would provide the sector with certainty and much-needed clarity. The only caveat is that, being still in its infancy, not all the space-active states have deliberated in this sense.

It has been observed (Marboe 2015) that there are both advantages and disadvantages legal norms, and, if they are directly applicable and enforceable, which is not necessarily the case with obligations of a public international law nature, they are also generally not applicable outside the national territory. They are limited in number, too, because to date "*national space legislation is available in only a few states. There are sufficient and forward-looking legal regimes in some countries, but this is not the case in others. This is an obstacle to the creation of a level playing field in the space sector. If all the actors were bound by similar and fair rules in the carrying out of space activities, this would be to the long-term benefit of all. By contrast, in a situation of many diverse national legal frameworks, the phenomena of 'nationality planning' or the choice of 'flags of convenience', as known from the law of the sea, are also potential problems relevant in the law of outer space*" (Marboe 2015: 45).

What follows does not aim at providing the full picture of the existing national laws in the space sector[33]—impossible for reasons of space—but to flag the most important (to date) existing laws and offer a few discussion points. An important exception is that everything concerning the EU as a transnational body (including ESA, even though the European Space Agency is not an EU institution itself) has been treated in Chapter 4, when discussing the European Space Policy as a whole.

7.5.1 USA[34]

U.S. Commercial Space Launch Competitiveness Act—Title 51 (Public Law 114–90 114th Congress, November 25, 2015)

Before the Obama Administration put a hand on it, there was no dedicated title in the U.S. Code for space programmes, albeit some of them have been later enacted to address parts of it. Title 51 included four sections, which cover private operators, commercial launches, commercial remote sensing and, more importantly, a ground-breaking title on the space resource commercial exploration and utilisation.

Art. 51 states "*a United States citizen engaged in commercial recovery of an asteroid resource or a space resource under this chapter shall be entitled to any asteroid resource or space resource obtained, including to possess, own, transport, use, and sell the asteroid resource or space resource obtained in accordance with applicable law, including the international obligations of the United States*" (Arts. 51–303).

Private companies looking for exploiting space resources—as explained in Chapter 8—received a major boost form the legislation, called "the single greatest recognition of property rights in history" by Eric Anderson[35] of Planetary Resources, one of the start-ups in the space mining business.

While granting basic legal protections for space mineral mining, however, the US law asserted no ownership on the celestial bodies themselves. The resources are for grab, but not the ground that contains them: under this aspect, Title 51 does not change 1967 OST provisions: it integrates them.

In addition to the provisions for the mining sector, the law also confirmed that "that the International Space Station remains a viable and productive facility capable of potential United States utilisation through at least September 30, 2024", also granting a "learning period" for commercial spaceflight ventures through 2023, with indemnification of commercial launches through 2025.

7.5.2 Luxembourg[36]

Loi du 20 juillet 2017 sur l'exploration et l'utilisation des ressources de l'espace. 20 July 2017

Based on Draft Law on the Exploration and Use of Space Resources, Ministry of Economy,

The Government of the Grand Duchy of Luxembourg, 11 November 2016,

It might come as a surprise to some people that this tiny state situated at the middle of the EU enacted a commercial space law so early in the space race. It is less of a surprise when one looks at the fundamental of the state itself. Luxembourg has long been the favourite location for company holdings in the EU; it has solid institutions, a strong banking sector, and host many international bodies. Moreover, the state has proven expertise and a track record in a segment of the space sector, namely the satellite business, since the 1980s.

When satellites stopped to the precinct of government-led efforts and opened to the private initiatives, Luxembourg led the way, (SES) Société Européenne des Satellites in 1985, Europe's first private satellite company and now the world's second-biggest commercial satellite operator.

The bid to the space mining, therefore, did not happen in the void.

Luxembourg drew up its first action plan for space research and development back in 2008 (Space Resource Luxembourg 2017), also setting up a structural framework, including R&D centres and university, for the business to develop.

"*Our goal is to put into place an overall framework for the exploration and commercial use of resources from 'celestial bodies' such as asteroids, or from the Moon*", Etienne Schneider, Luxembourg's Deputy Prime Minister and Minister of the Economy, declared.

In 2016, the state established the Space Resource Initiative and devoted US$223 million of space budget available to companies for feasibility studies. Luxembourg has already started working with two well-known enterprises in the sector, Deep Space Industries and Planetary Resources. In the second, the state has bought an equity share for the value of US$28 million.

When the space mining law passed in 2017, therefore, it was a surprise for nobody.

As for the content, the letter of the law mirrors the US one, when allowing the appropriation of the resources themselves (Art. 1[37]). However, the law goes even further, containing procedures for authorisation and supervision absent in the US legislation. The Grand Duchy subordinates the exploitation to a prior ministerial agreement, for each specific and determined mission (Arts. 2, 1[38]), and to the legal persons (not private citizens, just to be clear) either national or established there (Art. 4[39]).

7.5.3 Japan[40]

Basic Space Law (Law No. 43 of 2008). Promulgated on 28 May 2008

Soon after passing the legislation, Japan elaborated its Basic Plan for Space Policy (hereinafter, the "Basic Plan"), which clarifies the country's medium and long-term policy on the development and use of space.

The Basic Law overturned the previously existing legislation, 1969 Law[41] that, among other things, also instituted the Japanese space agency JAXA and confined the country's space development to peaceful purposes only. While the Basic Law still consider civilian and commercial objectives as paramount, strengthening the technical capabilities and international competitiveness of the space industry and other industries of Japan, (Art. 4),[42] development of Human Society (Art. 5) and international cooperation (Art. 6), also consider Japanese security needs (Art. 3), where explicitly declaring that the space development aims at "increasing the national security of Japan".[43]

An entire section is devoted to the satellite developing and launching (Chapter 2), with one specific article (Art. 16[44]) devoted to the Promotion of Space Development and Use by Private Business Operators. Under this point of view, Japan was one of the first countries to recognise and protect private operators.

Finally, in an attempt to unify Japan's space policies, the law also established the Strategic Headquarters for Space Policy (SHSP), composed of all ministers in the Cabinet.

Interestingly, the law made no mention of JAXA, which was therefore still bound to its 2002 law[45] and all-peaceful purposes. Probably spurred by the advance of China and India's space programme, (*Space News*, 30 January 2012, online), Japan quietly amended the Basic Space Law in 2012 based on the advice of the SHSP, passing in June 20, 2012, the Act on Partial Revision of the Act for Establishment of the Cabinet Office, (Act No. 43, 2012).

The Act (already mentioned in Chapter 5) was an important measure in administrative terms and set up a regime for promotion and use of space in accord with the recent-enacted Basic Space Law. Moreover, the Act revised Art. 4 of the JAXA's law in order to be consistent with the new principles indicated by the Basic Space Law, and that allows for military activities for defensive aims (as for 1967 OST). In Anan's words, "the re-definition of the peaceful purposes of JAXA was included in the bill. As a result of this amendment, like the JSDF, JAXA is expected to be able to operate its space programs based on the pacifist principles of Article 9 of the Japanese Constitution. In theory, this amendment enables JAXA to work as a core implementation agency to support the Ministry of Defence, with which JAXA could not cooperate before" (Anan 2013: 4).

7.5.4 The UK[46]

Outer Space Act 1986 ("OSA"), Chapter 38.

As one of the first states to engage in space activities, the UK also enacted in the 1980s a law that regulates the use of outer space. The law covered a few fundamental areas, such as the application procedures to obtain a licence (the so-called OSA Licence) required to launch or operate a space object (such as satellites) and to manage other activities in outer space.

The law was subjected to partial deregulation in 2015 (Deregulation Act 2015, Chapter 20[47]), which introduced limits to the operators' indemnity to the national government for third-party claims brought against the UK.

Mirroring the UK interests in space, which are more aimed at commercial spaceflights and operation of domestic spaceports and satellites rather than mining, the brand-new Space Industry Act 2018,[48] passed on 15 March 2018, regulates the first commercial space launches from the UK soil in history.

As the UK Transport Minister Jo Johnson declared "*access to space is the final piece of the puzzle for the UK space sector and forms part of the government's Industrial Strategy – helping to drive growth across the country and unlocking hundreds of highly-skilled jobs in our regions. The Space Industry Bill gives companies the ability to launch satellites from UK soil, putting us at the forefront of the new space race, and helping us to compete as the destination of choice for satellite companies worldwide*" (Jo Johnson,[49] as quoted on the UK Government's official website 2018, online).

7.5.5 China[50]

China's Space Activities in White Papers, 2000, 2006, 2011, 2016.

As strange as it might look, China has not (yet) a national law that regulates the access to space, either for launches or for mining resources, albeit both areas are of primary interest for the country. Moreover, China voiced no opposition or commented in any way to the US and Luxembourg's legislative initiatives that allow property rights on mined resources.

> This conduct is particularly relevant not only in the light of the growing importance of the Chinese space program but also if one considers that China is the country most involved in the exploration and study of celestial bodies and their resources, particularly the Moon. For this reason, it would have been legitimate to expect China to have a more engaged behaviour. However, China has acted otherwise. (Hao and Tronchetti 2017: 1)

The literature has offered two interpretative keys to this, the lack of a strong domestic private sector in the space sector (even though this is changing; see Chapter 3) and the fact that the legislative outlook in China follows different rules from Western countries and a different hierarchy of sources and enforcement procedures.

What exists at the moment is a series of White Papers on China's space activities (four to date), where China highlighted the purpose of its activities in space, and namely "*to explore outer space and enhance understanding of the Earth and the cosmos; to utilise outer space for peaceful purposes, promote human civilisation and social progress, and benefit the whole of mankind; to meet the demands of economic development, scientific and technological development, national security and social progress; to improve the scientific and cultural knowledge, protect national rights and interests, and build up overall strength*" (White Paper 2016, online).

It is worth to notice that, while China "*holds that all countries in the world have equal rights to peacefully explore, develop and utilise outer space and its celestial bodies, and that all countries' outer space activities should be beneficial to their economic development and social progress, and to the peace, security, survival and development of mankind*" (White Paper 2016, online), it makes no mention of property rights on the resource mined.

Furthermore, no mining endeavour is present in the plan for the next five years section of the latest white paper, not even as an area of future cooperation. This is likely to change sooner than later.

Case Study: Claiming Damages in Outer Space—The Case of Iridium 33-Cosmos 2251 Collision

That the staggering (and growing) amount of debris in the Earth's orbit was going to cause sooner or later an accident, everybody knew it since the beginning: that it has happened so rarely has been more surprising.

But of a few accidents that have occurred so far, one has attracted worldwide attention, for the wide array of implications and for the fact it involved governments and private actors, complicating the matter even further.

On 10 February 2009 the derelict, but still intact, Russian satellite Cosmos 2251[51] crashed into the active and privately-owned Iridium 33 satellite, which resulted in the destruction of the second, smaller than the first.[52] The collision happened in LEOs, at an altitude of 770 km (470 miles). It was the first time two satellites in orbit collided by accident, even though not first recorded accident between two artificial objects in space. The French military microsatellite Cerise[53] had been hit in 1996 by a known space debris coming from an Ariane rocket, and left seriously damaged by the impact, even though still able to continue its work afterwards. Not so the case of Iridium 33, which exploded in a myriad of yet more debris.

According to the U.S. Space Surveillance Network,[54] more than >10 cm 1800 debris (about 10% of the existing debris of the same size at that time) and a number far higher of smaller ones have resulted from that collision.

The legal dispute that followed was excruciating but also of interest for many reasons, because it highlighted the overwhelming challenges to apply the existing legislation, e.g., the 1972 Liability Convention that should have been able to address and solve them in a satisfactory way. Which was clearly not the case.

There were a few issues that made a solution to the case of extreme difficulty. The first and more important one was related to the possibility of collision avoidance.

This was not a menial point. Collision avoidance manoeuvres are performed every year. NASA started this practice in 2005 and, by 2007, a routine of conjunction assessments was required for all NASA's satellites in LEO and GEO. Up to 2009, when the accident occurred, many have been recorded, e.g., 10 by ISS alone, including twice in 2009, and 9 for the Space Shuttle when not docked to ISS. But, and this was a crucial point, the Russian satellite was not operational at that time: after being launched in 1993, it was reported to have ceased operation two years later. There was nothing the Russians could do to avoid the clash (short of destroying the satellite with a rocket. As the Chinese showed in 2007, not a brilliant idea either).

So the question rather was why Iridium did not do anything to manoeuvre its satellite away to avoid the collision. The answer is: because it did not expect it to happen. The problem is that the satellite tracking system, meant to help avoid these occurrences, is far from perfect. In this specific case, Celestrak's SOCRATES (Satellite Orbital Conjunction Reports Assessing Threatening Encounters in Space) had predicted a close approach of 584 m between Iridium 33 and Cosmos 2251 in about 14 reports during the week preceding the event, but none of them made the list of "Top 10 Conjunctions" by Maximum Probability or Minimum Range that are closely monitored at every moment.[55] This was one of the reasons why no corrective manoeuvre was attempted by Iridium, and the (miscalculated) approach instead resulted in a collision.[56]

Taking the above fact into consideration and looking at the Liability Convention's provisions, only the second scenario.[57] (where the fault liability applies only if it proved that the damage caused was due to the fault of the Launching State or States) was applicable. However, as Listner argued, "this theory of liability is similar to the maritime law standard of comparative fault, whereby the percentage of fault and thus liability will be apportioned according to the negligence of each party involved to the aggregate liability. Once the percentage of fault for each party is determined, the total amount of damages for the incident will be determined, and from that total each party will be apportioned its damages based on its percentage of fault. However, the conflicting information leading up to the crash coupled with the lack of tracking data for that orbital section of outer space gave neither party sufficient evidence to meet the fault standard set by the second scenario".

There are lessons to be learnt in the Iridium case, and the first one goes under the label "space sustainability"; cutting down all activities that result in the production of debris is probably the most important. One of the issues raised in a more concrete fashion by the Iridium-Cosmos accident was the ownership of the debris created by the impact. The major risk is caused not by larger debris, registered (and tracked) by US SNN, but by the smaller (and impossible to track) 200,000 debris with sizes between 1 and 10 cm. NASA's CloudSat satellite conducted just on 23 April 2009 one of the first collision avoidance manoeuvres to avoid debris produced by the Iridium 33–Cosmos 2251 collision, and others have followed. Now, for the Art. 8 of the UN-OST, they belong to the launcher states and only they have the authority for removal and destruction of them, while unilateral measures constitute a breach of sovereignty (Baker 1988; Christol 1990). This provision clashes against close cooperation among states to address and contain (if not solve) the growing problem of debris in orbit, like the recent Japanese initiative in this sense, and they have to be strongly encouraged.[58] A revision of this section of the treaty looks advisable, too, especially if considering that the majority of the commercial satellites in LEO are not insured after their first year of activity. And while collisions in outer space are not a common occurrence (yet), there's something that can be done in the sense of setting up an international control centre to coordinate satellite course (Lee-Volker 2007), using in better way data already existing. One is the already mentioned Celestrak-SOCRATES, whose functionalities have been improved since 2009.

What matters here is that even one more accident can raise the numbers of debris to a sky-high level (pun intended), if considered than just 11 of more than 4500 space missions flown since 1957 are responsible for 32% of all catalogued artificial Earth satellites presently in orbit according to recent NASA's estimates.[59] Otherwise said, the number of debris resulting from those events that have been significantly exceeding the theoretical models' estimates. Avoiding accidents in the Earth's orbit is rapidly becoming a number one priority, legal compensation notwithstanding.

Finally, the Iridium-Cosmos accident proved, if anything, that the existing 1972 LC required a set of pre-conditions to be effective, which were impossible to meet in the specific case, and arguably, in the majority of similar instances of the same kind arising in the future. Other and more sophisticated mechanisms of dispute resolution, involving both public and private actors and requiring less stringent prerequisites, have to be devised to avoid harsh legal battles in future years.

Sources

Celestrak-SOCRATES official tracking system: http://celestrak.com/SOCRATES/.

Jakhu, R. (2010). Iridium-Cosmos Collision and Its Implications for Space Operations. In *ESPI Yearbook on Space Policy. 2008/2009: Setting New Trends* (pp. 254–275). Wien and New York: Springer.

Kelso, T. S. (2012). *Analysis of the Iridium 33-Cosmos 2251 Collision*. Center for Space Standards and Innovation at http://celestrak.com/events/collision/.

Listner, M. (2012). Iridium 33 and Cosmos 2251, Three Years Later. *Space Safety Magazine*. Retrieved 12 January 2019, from http://www.spacesafetymagazine.com/space-debris/kessler-syndrome/iridium-33-cosmos-2251-years-later-learned-then/.

NASA. (2009). *The Collision of Iridium 33 and Cosmos 2251: The Shape of Things to Come*. https://ntrs.nasa.gov/archive/nasa/casi.ntrs.nasa.gov/20100002023.pdf.

Russell, P. P. (2001). General Method for Calculating Satellite Collision Probability. *Journal of Guidance Control and Dynamics, 24*, 716–722.

Wang, T. (2010). Analysis of Debris from the Collision of the Cosmos 2251 and the Iridium 33 Satellites. *Science & Global Security, 18*, 87–118.

Notes

1. The full list of agreements and regulations that regard space is longer than that, and it can be retrieved from the UNOOSA website, as discussed in the test. Here only the most important have been discussed.
2. The resolution is available at http://www.unoosa.org/oosa/SpaceLaw/gares/html/gares_14_1472.html.
3. All the details in this section have been directly taken from the official UNOOSA website, which has published a booklet with all the treaty laws and principles. UN. (2017). *International Space Law: United Nations Instruments*. Vienna: UN, available at http://www.unoosa.org/res/oosadoc/data/documents/2017/stspace/stspace61rev_2_0_html/V1605998-ENGLISH.pdf.
4. UNOOSA maintains a registry, which is updated every year. The latest one is A/AC.105/C.2/2018/CRP.3, Status of international agreements relating to activities in outer space as at 1 January 2018, available at http://www.unoosa.org/oosa/en/ourwork/spacelaw/treaties/status/index.html. The page also reports the latest signatures.
5. As of January 2018, only 18 states are parties of it.
6. Originally called "Treaty Banning Nuclear Weapon Tests in the Atmosphere, in Outer Space and Under Water" https://www.state.gov/t/isn/4797.htm.
7. For a discussion about the topic, see: https://www.nti.org/learn/treaties-and-regimes/proposed-prevention-arms-race-space-paros-treaty/ "The committee was dissolved in 1994, and it has yet to reconvene despite an annual, near-unanimous vote by the UN General Assembly (UNGA) and the CD approving the PAROS resolution. The United States is the only country to vote against the resolution, with Israel abstaining. The United States argues that PAROS is unnecessary because there are no weapons--and thus no arms race--in outer space at this time". The full text is available at https://fas.org/programs/ssp/nukes/ArmsControl_NEW/nonproliferation/NFZ/NP-NFZ-PAROS.html.
8. UNOOSA keeps track of all of them. The complete list is available on UNOOSA website: http://www.unoosa.org/pdf/misc/lts/Int._Agreements.pdf.
9. Art. 1, OST, 1967: "The exploration and use of outer space, including the Moon and other celestial bodies, shall be carried out for the benefit and in the interests of all countries, irrespective of their degree of economic or

scientific development, and shall be the province of all mankind. Outer space, including the Moon and other celestial bodies, shall be free for exploration and use by all States without discrimination of any kind, on a basis of equality and in accordance with international law, and there shall be free access to all areas of celestial bodies. There shall be freedom of scientific investigation in outer space, including the Moon and other celestial bodies, and States shall facilitate and encourage international cooperation in such investigation".

10. Art. 2, OST, 1967: "Outer space, including the Moon and other celestial bodies, is not subject to national appropriation by claim of sovereignty, by means of use or occupation, or by any other means".
11. Art. 6, OST, 1967: "States Parties to the Treaty shall bear international responsibility for national activities in outer space, including the Moon and other celestial bodies, whether such activities are carried on by governmental agencies or by non-governmental entities, and for assuring that national activities are carried out in conformity with the provisions set forth in the present Treaty. The activities of nongovernmental entities in outer space, including the Moon and other celestial bodies, shall require authorization and continuing supervision by the appropriate State Party to the Treaty. When activities are carried on in outer space, 6 including the Moon and other celestial bodies, by an international organization, responsibility for compliance with this Treaty shall be borne both by the international organization and by the States Parties to the Treaty participating in such organization".
12. Art. 8, OST, 1967: "A State Party to the Treaty on whose registry an object launched into outer space is carried shall retain jurisdiction and control over such object, and over any personnel thereof, while in outer space or on a celestial body. Ownership of objects launched into outer space, including objects landed or constructed on a celestial body, and of their component parts, is not affected by their presence in outer space or on a celestial body or by their return to the Earth. Such objects or component parts found beyond the limits of the State Party to the Treaty on whose registry they are carried shall be returned to that State Party, which shall, upon request, furnish identifying data prior to their return".
13. Art. 9, OST, 1967: "In the exploration and use of outer space, including the Moon and other celestial bodies, States Parties to the Treaty shall be guided by the principle of cooperation and mutual assistance and shall conduct all their activities in outer space, including the Moon and other celestial bodies, with due regard to the corresponding interests of all other

States Parties to the Treaty. States Parties to the Treaty shall pursue studies of outer space, including the Moon and other celestial bodies, and conduct exploration of them so as to avoid their harmful contamination and also adverse changes in the environment of the Earth resulting from the introduction of extraterrestrial matter and, where necessary, shall adopt appropriate measures for this purpose. If a State Party to the Treaty has reason to believe that an activity or experiment planned by it or its nationals in outer space, including the Moon and other celestial bodies, would cause 7 potentially harmful interference with activities of other States Parties in the peaceful exploration and use of outer space, including the Moon and other celestial bodies, it shall undertake appropriate international consultations before proceeding with any such activity or experiment. A State Party to the Treaty which has reason to believe that an activity or experiment planned by another State Party in outer space, including the Moon and other celestial bodies, would cause potentially harmful interference with activities in the peaceful exploration and use of outer space, including the Moon and other celestial bodies, may request consultation concerning the activity or experiment".

14. "The primary strategy for preventing contamination of Mars with Earth organisms is to be sure that the hardware intended to reach the planet is clean. [...]Another type of precaution is to be sure that other hardware doesn't go to Mars accidentally. When the Delta launch vehicle's third stage separated from the spacecraft, the two objects were traveling on nearly identical trajectories. To prevent the possibility of the third stage hitting Mars, that shared course was deliberately set so that the spacecraft would miss Mars if not for its first trajectory correction manoeuver, about 10 days later" (NASA 2018, online).
15. Interestingly, the later inclusion of this specific provision was due to an Air Force project and its clashes with the international scientific community (especially the IAI) Terrill (1999).
16. For the way this requisite applies to ASATs and dual-use object, see Chapter 6.
17. The treaty also recommends the institution of "*an international regime should be established to govern the exploitation of such resources when such exploitation is about to become feasible*" (Art. 11).
18. At this purpose, in space law literature this is the definition of dispute that is current. ""A specific disagreement concerning a matter of fact, law or policy in which a claim or assertion of one party is met with refusal,

counter-claim or denial by another" (Merrils 1998: 1). It must include therefore two or more conflicting claims over the same object.

19. When more states are involved, for example one for the launching facility and the other for the territory, they are going to be "jointly and severally liable for any damage caused" (Art. 5).
20. Art. 3, LC, 1972: "In the event of damage being caused elsewhere than on the surface of the Earth to a space object of one launching State or to persons or property on board such a space object by a space object of another launching State, the latter shall be liable only if the damage is due to its fault or the fault of persons for whom it is responsible".
21. Art. 10, LC, 1972: "A claim for compensation for damage may be presented to a launching State not later than one year following the date of the occurrence of the damage or the identification of the launching State which is liable".
22. Art. 10, LC, 1972: "The Claims Commission shall be composed of three members: one appointed by the claimant State, one appointed by the launching State and the third member, the Chairman, to be chosen by both parties jointly. Each party shall make its appointment within two months of the request for the establishment of the Claims Commission. 2. If no agreement is reached on the choice of the Chairman within four months of the request for the establishment of the Commission, either party may request the Secretary-General of the United Nations to appoint the Chairman within a further period of two months".
23. Art. 1, LC, 1972: "The term 'damage' means loss of life, personal injury or other impairment of health; or loss of or damage to property of States or of persons, natural or juridical, or property of international intergovernmental organizations".
24. It is the case to observe that, together with these procedures at international level, there are some other recourses offered by the various international organisations involved at any title in the space activities. One example is ESA and its arbitration.Procedures (see Chapter 4) or ITU, in the case of harmful interference to registered frequencies or similar cases (Jakhu 2013). Also, there is always possible appealing to the ICJ for any disputes at state level but again, as mentioned for 1972 CL, this is not open to private citizen, but only to states.
25. The full text is available at https://pca-cpa.org/wp-content/uploads/sites/175/2016/01/Permanent-Court-of-Arbitration-Optional-Rules-for-Arbitration-of-Disputes-Relating-to-Outer-Space-Activities.pdf.

26. All OECD countries among them, but also China and Russia. Full list available at https://pca-cpa.org/en/about/introduction/contracting-parties/.
27. See Chapter 4 for details about disputes with China and the US on this specific point.
28. US launchers privilege larger payloads as primary and the small satellites need to adapt.
29. This is Art. XVI, GATS. For the full text of GATS, General Agreement on Trade in Services, see WTO's official website, https://www.wto.org/english/docs_e/legal_e/26-gats_01_e.htm.
30. This is not so strange when thinking only a few specific recourses exist even for state-to-state negotiations as previewed by the 1972 LC, as specified in the previous section.
31. See for example, the documents of some of the latest sessions, at http://www.unoosa.org/oosa/en/ourwork/copuos/comm-subcomms.html.
32. The Hague Space Resources Governance Working Group is composed by stakeholders from government, industry, universities, civil society and research centres from the international space community. See more here: https://www.universiteitleiden.nl/en/law/institute-of-public-law/institute-for-air-space-law/the-hague-space-resources-governance-working-group.
33. There is a growing corpus of documents and literature in this sense. Here the book has adopted the original list resident on UNOOSA website: http://www.unoosa.org/oosa/en/ourwork/spacelaw/nationalspacelaw.html, integrated with the national repositories of the countries mentioned in the section.
34. For the official text of Title 51, see: https://www.gpo.gov/fdsys/pkg/uscode-2010-title51/pdf/uscode-2010-title51.pdf.
35. Jamasmie, C. (2015). *US Congress Grants Citizen Rights to Own Asteroid-Mined Riches* |MINING.com. Retrieved 4 December 2018 from http://www.mining.com/us-congress-grants-citizens-rights-to-own-asteroid-mined-riches/.
36. The text of the law is available at http://legilux.public.lu/eli/etat/leg/loi/2017/07/20/a674/jo.
37. Art. 1: "Les ressources de l'espace sont susceptibles d'appropriation". It translates into "space resources are capable of being owned", therefore giving legal permission to space mining.
38. Art 2: "Aucune personne ne peut explorer ou utiliser les ressources de l'espace sans être en possession d'un agrément de mission écrit du ou

des ministres ayant dans leurs attributions l'économie et les activités de l'espace".

39. Art. 4: "L'agrément pour une mission ne peut être accordé que si le demandeur est une société anonyme, une société en commandite par actions ou une société à responsabilité limitée de droit luxembourgeois ou une société européenne ayant son siège social au Luxembourg". For more about Luxembourg and its laws in the space sector, see Luxinnovation (2018).
40. A good starting point for Japanese laws and policies for outer space is on the English section of JAXA website at http://global.jaxa.jp/about/law/an. The English translation of the Japanese text is available at http://stage.tksc.jaxa.jp/spacelaw/country/japan/27a-1.e.pdf.
41. Law Concerning the National Space Development Agency of Japan (Law No. 50 of June 23, 1969, as amended) available at http://www.unoosa.org/oosa/en/ourwork/spacelaw/nationalspacelaw/japan/nasda_1969E.html.
42. Art. 4:"(Advancement of Industries) Space Development and Use shall be carried out in order to strengthen the technical capabilities and international competitiveness of the space industry and other industries of Japan, thereby contributing to the advancement of the industries of Japan, by the positive and systematic promotion of Space Development and Use as well as smooth privatization of the results of the research and development with regard to Space Development and Use".
43. It is not a coincidence that the Basic Law appeared one year after China blew up a weather satellite in space (in January 2007); after that, Japan also launched its own recon satellites for military purposes.
44. Article 16: "Taking into account the important role of private operators in Space Development and Use, and in order to encourage business activities (including research and development activities) with respect to Space Development and Use in the private sector, aiming at strengthening the technical capabilities and international competitiveness of space industry and other industries of Japan, the State shall, in conducting its own Space Development and Use, consider the procurement of goods and services systematically using the capabilities of private operators, the improvement of launching sites (meaning installations for the launching of rockets), experiments and research facilities as well as other facilities and installations, the promotion of the transfer of results of the research and development with regard to Space Development and Use to private business operators, the promotion of the privatization of the results of

research and development with regard to Space Development and Use in the private sector, and the taking of taxation and financial measures and other necessary measures in order to facilitate investment by private operators in the business with regard to Space Development and Use".

45. Law concerning Japan Aerospace Exploration Agency (Law Number 161 of 13th December 2002) http://global.jaxa.jp/about/law/law_e.pdf.
46. All references are from the UK Gov website. For the OSA Text, see https://www.Legislation.gov.uk/ukpga/1986/38/pdfs/ukpga_19860038_en.pdf.
47. Available at http://www.legislation.gov.uk/ukpga/2015/20/pdfs/ukpga_20150020_en.pdf.
48. For more details about the parliament discussions and the approval process of the Bill, see https://services.parliament.uk/bills/2017-19/spaceindustrybill.html.
49. The whole speech is available at https://www.gov.uk/government/news/new-laws-unlock-exciting-space-era-for-uk.
50. All the full texts of China's White Paper are available online. White Paper 2000: http://www.china.org.cn/e-white/8/index.htm. White Paper 2006: http://www.gov.cn/english/2006-10/12/content_410941.htm. White Paper 2016: http://english.gov.cn/archive/white_paper/2016/12/28/content_281475527159496.htm.
51. In the Russian use, all spacecraft and satellites not recovered or otherwise lost are termed "Cosmos", no matter their initial name.
52. The Russian satellite had a mass of about 900 km and it was intact but not active and therefore not manoeuvrable any longer at the time of the accident. Iridium 33 was smaller, about 556 kg part of the constellation of commercial satellites owned by private US company Iridium Satellite LLC.
53. Cerise was weighting only 50 kg and had been launched one year before to eavesdrop HF communications. Albeit the impact, it was reprogrammed and continued its work.
    More about the accident here at: https://nssdc.gsfc.nasa.gov/nmc/spacecraft/display.action?id=1995-033B.
54. As of 2012, June 22, the SSN has already catalogued 598 pieces of debris (93 pieces of which already decayed from their orbit) of Iridium 33 and 1603 pieces of debris (222 decayed) from Cosmos 2251.
55. See for instance a daily list here: http://celestrak.com/SOCRATES/.

56. The report and the analysis of the tracking (Kelso 2012) can be analysed by using the STK Viewer 9 available on Celestrak's website at http://celestrak.com/events/collision/#Viewer.
57. The first scenario is absolute liability. Another issue that soon emerged is that it was unclear which one was to be considered the launching state of Iridium 33 It was owned by a US company, launched by a Russian vector (Proton) from a Kazakh spaceport, Baikonur. Moreover, the satellite was not even registered with the UN, as required by the 1974 Registration Convention (Ram 2010).
58. And so it should be with all the mechanisms that share orbital data with all the interested parts, on the kind of agreement Russia and the USA already have with the Joint Space Operations Centre (JSpOC), which was quite helpful to Russia in the case of its probe Phobos-Grunt in December 2011.
59. There are many theoretical models that try to estimate the amount of debris produced by break up and collision in Earth's orbit, the most used one being NASA's StandardBreakup Model, an empirical model based on historical occurrences and ad hoc testing. "The model gives estimates of distributions of fragment parameters based on the collision velocity and mass of the colliding objects" (Wang 2010: 3). It does not come as a surprise that the models are often not accurate in their prediction and should only been taken as a guidance. For instance, NASA's model underestimated the break up resulting by the other major event in terms of debris formation, the (intentional) destruction of the Chinese satellite Fengyun in January 2007 by a factor of three (Russell 2001).

References

Anan, K. (2013). Administrative Reform of Japanese Space Policy Structures in 2012. *Space Policy*. Available at https://doi.org/10.1016/j.spacepol.2013.06.001.

Baker, H. A. (1988). Liability for Damage Caused in Outer Space by Space Refuse. *Annals of Air & Space Law, 8*, 183–227.

Burnett, D. J., & Lihani, D. (1997). U.S. National Space Policy and Bilateral Launch Service Agreements. In *Proceedings of the Thirty-Ninth Colloquium on the Law of Outer Space*, 263–270.

Burnett, D. J., & Schroeder, F. O. (1994). "Development in US Bilateral Launch Service" Agreements. *Air & Space Law, 19*, 326–327.

Chatterjee, P. (2014). Legality of Anti-Satellite Under the Space Law Regime. *Astropolitics, 12*, 27–45.

Christol, C. Q. (1990). Suggestions for Legal Measures and Instruments for Dealing with Debris. *Environmental Aspects of Activities in Outer Space: State of the Law and Measures of Protection, 13*, 268–280.

De Selding, P. B. (2016, March 29). U.S. Launch Companies Lobby to Maintain Ban on Use of Indian Rocket. *Space News.* Available at http://spacenews.com/u-s-space-transport-companies-lobby-to-maintain-ban-on-use-of-indian-rockets. Retrieved on 18 December 2018.

Freedland, S., & Blake, D. (2017, July 9). *As the World Embraces Space, the 50 Year Old Outer Space Treaty Needs Adaptation.* The Conversation. Available at https://theconversation.com/as-the-world-embraces-space-the-50-year-old-outer-space-treaty-needs-adaptation-79833. Retrieved on 3 December 2018.

Gorove, S. (1973). Arms Control Provisions in the Outer Space Treaty: A Scrutinizing Reappraisal. *Georgia Journal of International and Comparative Law, 3*, 114–125.

Gorove, S. (1991). *Developments in Space Law: Issues and Policies* (pp. 239–241). Dordrecht: Kluwer.

Hansson, A., & McGuire, S. (1999). Commercial Space and International Trade Rules: An Assessment of the WTO's Influence on the Sector. *Space Policy, 15*, 203–204.

Hao, L., & Tronchetti, F. (2017). Should the Red Dragon Arise? Assessing China's Options Vis-a-Vis the Enactment of a Domestic Space Resources Utilisation Law. *Space Policy, 39–40*, 9–13.

He, Q. (1991). Legal Issues of China's Entry into International Space Market. *Zeitschrift für Luft- und Weltraumrecht, 40*, 280–291.

Hobe, S. (2007). Adequacy of the Current Legal and Regulatory Framework Relating to the Extraction and Appropriation of Natural Resources. *Annals of Air and Space Law, 32*, 114–135.

Horsley, N. (2005). The Arianespace Monopoly, EU Competition Law, and the Structure of Future European Launch Markets. *Air & Space Law, 30*, 87–95.

Jackson, J. H. (1969). *World Trade and the Law of the GATT.* New York: Bobbs Merill Co.

Jakhu, R. (2013). *Dispute Resolution Under the ITU Agreements.* Available at http://swfound.org/media/48115/Jakhu-Dispute%20resolution%20under%20the%20ITU%20agreements.pdf.

Lai, B. C. (1993). National Subsidies in the International Launch Market. *Space Policy, 9,* 23–25.

Listner, M. J. (2011, October 24). *The Moon Treaty: Failed International Law or Waiting in the Shadows?* The Space Review. Available at http://www.thespacereview.com/article/1954/1. Retrieved on 18 December 2018.

Listner, M. J. (2016). India's Commercial Space Conundrum. *Space Thoughts.* Available at https://spacethoughtsblog.wordpress.com/2016/07/06/indias-commercial-space-conundrum/. Retrieved on 10 January 2017.

Luxinnovation. (2018). *Luxembourg Space Cluster Website.* Available at https://www.luxinnovation.lu/cluster/luxembourg-space-cluster/. Retrieved on 11 February 2018.

Marboe, I. (2015). *National Space Law in Handbook of Space Law.* London: Elgar Elwin.

McCormick, P. K., & Mechanick, M. J. (2013). The *Transformation of Intergovernmental Satellite Organisations.* Leiden: Martinus Nijhoff Publishers.

Merrils, J. G. (1998). *International Dispute Settlement.* Cambridge: Cambridge University Press.

Mineiro, M. (2008). FY-1C and USA-193 ASAT Intercepts: An Assessment of Legal Obligations Under Article 9 of the Outer Space Treaty. *Journal of Space Law, 34,* 315–325.

Minnesota Journal of International Law. (2017, April 7). Including Launch Services in the GATS Can Relieve the Drought in the United States Small Satellite Launch Market. Available at http://minnjil.org/including-launch-services-in-the-gats-can-relieve-the-drought-in-the-united-states-small-satellite-launch-market/#_ftn14. Retrieved on 4 December 2018.

NASA. (2018). Available at https://mars.nasa.gov/mer/technology/is_planetary_protection.html.

ONU. (2017). *United Nation Report.* Available at https://www.un.org/development/desa/en. Retrieved on 10 February 2018.

PCA. (2018). *Permanent Court of Arbitration.* Available at https://pca-cpa.org.

Pike, G. (1991). Chinese Launch Services: A User's Guide. *Space Policy, 7,* 103–115.

Scarborough, J. (1992). Free Trade and the Commercial Launch Industry. *Space Policy, 8,* 109–120.

Scott, R. W. (1996). The Challenge to Commercial Space Transportation in the 21st Century. *Journal of Space Law, 24,* 152–157.

Singh, N. K., Bezdan, D., Checinska Sielaff, A., Wheeler, K., Mason, C. E., & Venkateswaran, K. (2018). Multi-drug Resistant Enterobacter Bugandensis Species Isolated from the International Space Station and Comparative Genomic Analyses with Human Pathogenic Strains. *BMC Microbiology, 18*(1), 175.

Space News. (2012, January 30). *Japan Moves To Relax Restrictions on Military Space Development.* Available at https://spacenews.com/japan-moves-relax-restrictions-military-space-development. Retrieved on 5 December 2018.

Space Resource Luxembourg. (2017). *Luxembourg Space Capabilities—Luxembourg Cluster Catalogue 2017.* Available at http://clustermembers.luxinnovation.lu/space/wp-content/uploads/sites/4/2016/10/07697_LUXINNOVATION_SpaceCapabilities_05-2017-Web.pdf. Retrieved on 10 January 2017.

The Conversation. (2018). *Mining Asteroids Could Unlock Untold Wealth—Here's How to Get Started* (May 2, 7:25 a.m. EDT). Available at http://theconversation.com/mining-asteroids-could-unlock-untold-wealth-hereshow-to-get-started-95675. Retrieved on 29 November 2018.

Terrill, Delbert R., Jr. (1999, May). Project West Ford. In *The Air Force Role in Developing International Outer Space Law. Air Force History and Museums* (pp. 63–67). Maxwell Air Force Base, AL: Air University Press.

Tronchetti, F. (2009). *The Exploitation of Natural Resources of the Moon and Other Celestial Bodies: A Proposal for a Legal Regime X.* Leiden: Martinus Nijhoff Publishers.

Twigg, J. L. (1994). The Russian Space Programme: What Lies Ahead? *Space Policy, 10,* 19–31.

UN. (2017). *International Space Law: United Nations Instruments.* Vienna: UN. Available at http://www.unoosa.org/res/oosadoc/data/documents/2017/stspace/stspace61rev_2_0_html/V1605998-ENGLISH.pdf. Retrieved on 18 December 2018.

UNOOSA. (2018). Official Website. http://www.unoosa.org/. Retrieved on 6 December 2018.

von der Dunk, F. (2015). International Trade Aspects of Space Services. In F. G. Von der Dunk & F. Tronchetti (Eds.), *Handbook on Space Law.* London: Edward Elgar.

Walter, C. E. (2011). The Privatisation and Commercialisation of Outer Space. In C. Brünner & A. Soucek (Eds.), *Outer Space in Society, Politics and Law* (pp. 494–506). Wien, New York: Springer.

WSJ. (2017, February 15). *India Breaks Record for Launching Most Satellites from a Single Rocket.* Available at http://blogs.wsj.com/indiarealtime/2017/

02/15/india-breaks-record-for-launching-most-satellites-from-single-rocket/. Retrieved on 07 February 2018.

WTO. (2017). *General Agreement on Trade in Services.* WTO.org, Art. XVI. Available at https://www.wto.org/english/docs_e/legal_e/26-gats_01_e.htm. Retrieved on 6 March 2017.

Zaborskiy, V. (2006). Space Engagement with Russia and Ukraine: Preventing Conflicts and Proliferation. *Astropolitics, 4,* 179–195.

8 A Changing Space: Trends of the Twenty-First Century

8.1 The Present and Future Telescope

If it is true that space exploration started when humankind looked for the first time at a moonless night trying to figure out stars and designing constellations, for the first conscious effort to study it in a more systematic and scientific way, the world needed to wait until Galileo and its prototype telescope, as simple as it was.[1]

Centuries have passed by and now the Earth is covered by watchful, artificial eyes peering into the sky, located in various areas of the planet. A few of them are rightly famous.[2]

The world's biggest telescope is, since 2009, the Gran Telescopio CANARIAS (GTC), a 34-foot (10.4-metre) giant located on La Palma of Spain's Canary Islands. It houses many support instruments such as CanariCam, a camera using mid-range infrared light emitted by stars and planets for in-depth analysis.

Hawaii's Mauna Kea Twin Telescopes (Keck I and II) are rightly famous. Their 33-foot (10-metre) mirrors are, after all, the second largest optical telescopes on the Earth's surface, in operation since the 1990s.

S. Paladini, *The New Frontiers of Space*,
https://doi.org/10.1007/978-3-030-19941-8_8

SALT, in Sutherland, South Africa, with its 30-foot (9.2-metre) telescope is the largest ground-based optical instrument in the Southern Hemisphere since it began operation in 2005. Moreover, it also "deployed the first laser guide star adaptive optics system on a large telescope in 2004, which creates an artificial star spot as a reference point to correct for atmospheric distortions when viewing the sky" (Space.com 2011, online).

However, what is probably the most famous observatory in the world is Arecibo, built in Puerto Rico in 1963. Arecibo's main feature is a 1000-foot (305-metre) radio reflector dish, which makes it the biggest radio telescope in the world, capable of capturing incredibly distant radio sources, such as quasars and remote galaxies.

ESO's radio telescopes, all located in Chile's Atacama Desert, are also well known to the astronomers. Atacama Large Millimetre/submillimetre Array (ALMA), while smaller than Arecibo in maximum diameter, is capable of better results because of its settings: the main array is made of 66, 39-foot (12-metre) high-precision radio antennas. Since its inception in 2013, ALMA[3] has been "the most powerful telescope for observing the cool Universe—molecular gas and dust" (ESO-ALMA 2019, online).

It is common knowledge that telescopes on the Earth, even powerful exemplars as the ones mentioned above, are limited by the objective conditions of the planet. The atmosphere that shelters the Earth from cosmic rays and solar radiations also reduces a great deal the visibility. Until the 1980s the trend has been to enhance the size of the primary mirror, to compensate for the reduced visibility. Intuitively, size does matter, because the larger the mirror, the more light it can collect, and the better resolution it can deliver.

The breakthrough came with the space telescopes in the 1980s, whose first examples were Copernicus Orbiting Astronomical Observatory, International Ultraviolet Explorer (IUE) and NASA's UHURU mission (SAS-A, launched in 1970), which allowed for an unparalleled, clear view of the celestial objects and demonstrated the interest to build a full-fledged space telescope: Hubble.

A joint ESA/NASA project, launched into Earth's orbit on 24 April 1990, the space telescope Hubble has dramatically changed our knowledge of the universe, making possible observations otherwise not feasible from

Earth. Hubble detects ultraviolet and infrared emissions and otherwise blocked by the atmosphere, receiving neat images.[4]

After Hubble's amazing science, other space telescopes followed.

Launched back in 1999, Chandra[5] is NASA's flagship mission for X-ray astronomy, with the task of detecting X-ray emissions. These waves from "very hot regions of the Universe such as exploded stars, clusters of galaxies, and matter around black holes. Because X-rays are absorbed by Earth's atmosphere, Chandra must orbit above it, up to an altitude of 139,000 km (86,500 miles) in space" (Chandra-X Ray Observatory 2018, online). Planet Hunter Kepler, another NASA's space telescope, has just been retired in 2018.[6] It was the end of a glorious mission: Kepler exoplanet-hunting discoveries have stunned the scientists in terms of data delivered and discovers achieved, discussed in Chapter 9.

New, powerful orbiting eyes are about to be put at work: Hubble's successor is the James Webb Space Telescope (JWST),[7] due to launch in 2021.

Hubble size was somehow reduced compared to the Earth telescope. It's only (so to speak) 2.4 metres in diameter, neatly folded into the rocket that put it in orbit, but JWST will be 6.5 metres in diameter. Moreover, JWST will be optimised for infrared wavelengths and will be more sensitive than Hubble, covering longer wavelengths of light. Thanks to its lower wavelength observation capabilities, JWST will be able to look further back in time to see the first galaxies, which formed during the early universe moments, also peering inside dust clouds where stars and planetary systems are still in formation.

Interestingly, the future of astronomy resides in the extremely big and extremely small (Wild 2016, 2017).

The next-generation ground telescopes belong to what is called "Extremely Large Telescope" (ELT). There are three, currently in construction, in the 98-foot (30-metre) range or bigger, which would house huge mirrors for collecting light from distant cosmic objects.

One is being built by ESO since 2005, and it aims at "vastly advance astrophysical knowledge, allowing detailed studies of subjects including planets around other stars, the first objects in the Universe, super-massive black holes, and the nature and distribution of the dark matter and dark energy which dominate the Universe" (ESO 2019, online).

ESO's telescope, located, like its smaller siblings, in the Atacama Desert in northern Chile, will have a 39-metre main mirror,[8] and it will be (at least for a while) the largest optical/near-infrared telescope on Earth, "the world's biggest eye on the sky".

The second, now in construction in Mauna Kea, Hawaii, is the so-called Thirty Meter Telescope[9] (TMT 2018, online), a US$1.4 billion telescope of 98 foot (30-metre) of aperture, due to enter into service in 2025–2030.

Finally, another X-ray telescope, ATHENA,[10] has been selected as the second European Space Agency (ESA) Large mission (L2). It has an intended launch date of 2028.

Additional initiatives are just behind the corners.

In a clear sign of the mutated geopolitical conditions of the space race, now it is not only space agencies like NASA or ESA to put space telescopes in orbit. China is getting there, too. In June 2017, China launched Hard X-ray Modulation Telescope (HXMT), an X-ray astronomical satellite and the first of its kind of Chinese production. While the telescope has only completed its in-orbit tests in February 2018, it has already proved its usefulness: it was one of the satellites that contributed to the detection of long-predicted and never-proved (until that moment) gravitational waves, one of the main discoveries of 2017.[11] China has also declared its intention to launch a space telescope akin to the Hubble Space Telescope in the mid-2020s, linked to China's Space Station, which will be there to offer permanent, operative support.

There are evolutions in other senses as well, and they go in the sense of the already noticed growing popularity and participation of the civil society—including universities—to the collective efforts of space observation and exploration. Telescopes is yet another field this trend can be witnessed.

Space telescope diameters are limited by the size of rockets they have to launch them to space. In 2008, students at Keck Institute of Space Studies (KISS) at Caltech showed that a new, modular design was the way to go and that it was possible to organise to send to the Earth's orbit multiple pieces that would self-assemble into a working telescope. The result was Autonomous Assembly of a Reconfigurable Space Telescope (AAReST), the brainchild of cooperation among Caltech's Ae 105 Aerospace Engineering student class, Surrey Space Centre in England, and the Indian

Institute of Space Science and Technology. It was completed in 2018 and AAReST satellite is scheduled to be launched[12] by the Indian Space Research Organisation (ISRO) rocket in 2019.

The most important thing to keep in mind is that, since there's a close linkage between space mission capabilities and space observatories, the development of reusable vectors inaugurated by SpaceX is going to affect the viability of the latter as well. "For most major astronomical observatories in space, the spacecraft is about half the cost. Cutting spacecraft cost by a factor 2, along with launch costs by a factor 3, would roughly halve the cost of a flagship mission" (Elvis 2017: 12).

With the development of new, more affordable highlighted in Chapter 3, there are high hopes of a new era even in terms of space telescopes.

8.2 Mining in Space

How does it cost mining an asteroid? This deceptively simple question is the key to viable exploitation of raw materials from outer space, be it from the ones orbiting the Earth (the so-called NEOs, near-Earth objects), the Moon, visiting, spent comets, or celestial bodies farther away. Even if the missions keep at what is close to Earth, the numbers are staggering: according to ESA (2017), of about 600,000 identified asteroids in our Solar System, more than 16,000 are NEOs. The destination of use is another important point, which will be covered later in this section.

First of all, any serious analysis should investigate the feasibility first. There have been a few studies in this sense. In 2012, KISS[13] sponsored a feasibility study for an Asteroid Redirect Mission (ARM), aimed at the capture of an asteroid to bring it in into lunar orbit. Among the participants of the study, there was also the CEO of Planetary Resources, one of the companies that are planning space mining. The study went into extreme details what such a mission would entail in terms of technology and even went into cost estimates for an asteroid of 23 feet (7 metres) wide in NEO and the weight of 500 tons, to be set in lunar orbit: US$2.6 billion (in USD of 2012).

Among the prerequisites of the mission, there are three conditions (Space.com 2012a, online): “Development of a ground-based observation campaign to discover and characterise potential target asteroids for return; development of a sufficiently powerful solar electric propulsion system necessary to transport the asteroid back to lunar orbit in a reasonable time Establishment of a human presence in lunar orbit to investigate and exploit this resource.”

Based on those conclusions, NASA proposed in 2013 a mission, The ARM[14] (AKA the Asteroid Initiative), aimed at visiting a large NEOs, collecting samples from its surface, putting into a boulder to be left orbiting around the moon, where the astronauts would explore and return to Earth with samples at some moment in the 2020s.

NASA has extensively worked on this area, identifying “multiple candidate asteroids and continues the search for one that could be redirected to near the moon in the 2020s. Since the announcement of the Asteroid Initiative in 2013, NASA’s Near-Earth Object Observation Program has catalogued more than 1,000 new near-Earth asteroids discovered by various search teams. Of those identified so far, four could be good candidates for ARM” (NASA-ARM 2018, online).

After many discussion in one sense or another—due to the scepticism in the Congress about its viability (*SpaceNews*, 14 June 2017), the mission was finally cancelled under White House Space Policy Directive 1, issued 11 December 2017 by President Trump, Foust (2017) which redefined the objectives of NASA for the current administration and that has been analysed in detail in the case study of Chapter 5.

While this specific mission was taken off the board, it is still likely to have other initiates, maybe crowdfunding or PPP as such. There is a good reason for that: they can advance a perspective Mars colony, testing the capabilities needed for a crewed mission in the 2030 and allow them to be no longer “Earth Reliant” like the International Space Station (ISS). Planetary Resources plans to go further than a robotic arm, though, and aims at building a “swarms of low-cost robotic spacecraft to extract resources from NEOs” using 3-D printing technology.

Another good question is what is to mine from space.

Precious metals seem the most logical answer (especially the platinum-group metals, which are found in such low concentrations on Earth’s soil—

such as rhodium, palladium, osmium, ruthenium, rhodium, palladium, osmium, iridium and platinum). After all, asteroids are small rocky worlds full of them, and "a single platinum-rich space rock 1,650 feet (500 metres) wide contains the equivalent of all the platinum-group metals ever mined throughout human history" (Space.com, 23 April 2012b).

Some scholars even estimated how big an asteroid should be to make it worth mining it at all. Elvis (2014) put this at US$1 billion in terms of resources. Other researchers made even more accurate calculations in terms of asteroid mining assets and the cost of retrieving them, as showed by Mainzer et al. (2011) "Assuming for now that only Ni–Fe asteroids are of interest for PGMs, then 1% of NEOs are rich in PGMs. The dearth of ultra-low delta-v (<4.5 km s−1) NEOs larger than 100 m diameter reduces the ore-bearing fraction to only ~1 in 2000 NEOs. As 100 m diameter NEOs are needed to have a value ≥US$ 1B and the population of near-Earth objects (NEOs) larger than 100 m diameter is ~20,000". However when weighting in the cost to reach them, the number of making profitable to mine them go down rapidly (down to 10) (Mainzer et al. 2011, online).

It is also evident that any improvement in the propulsion system will also increase.

Other, less evident raw material is water. According to Elvis estimates, when mining water is considered valuable asteroids number go as high as 9000. While common, water can be still very worth collecting in space, as it can be used to obtain oxygen and hydrogen for rockets.

Finally, Helium 3, also called H3D. In a future of nuclear fusion reactors, which looks likely at some moment in time, H3D will be priceless to fuel those reactors. And, while the element is painfully scarce on the Earth's soil, there is an abundance of it on the regolith [water could also be mined from the Moon soil. See Blair et al. (2002)]. China has shown interest in mining the lunar soil to extract H3D and there are missions already planned ahead. Another important aspect is where those space minerals are going to be used for.

While some of the rarest materials could be shipped back to Earth, the KISS study highlighted the importance of in situ utilisation, especially in the case of exploitation in lunar orbit of the several multi-hundred-ton asteroids that could be set in there, thus stimulating "the expansion of international cooperation in space as agencies come together to determine how to sample and process raw material from an asteroid. Water or other material extracted from a captured volatile-rich near-Earth asteroid could be used to provide affordable spacecraft shielding against galactic cosmic rays. The extracted water could also be used for propellant to transport a shielded habitat" (KISS 2012: 30).

Technical challenges to space mining are immense, and yet there is something trickier but not less critical in terms of viability: the legal framework. Who has the right to access those resources? The international treaties—examined in Chapter 7 in detail—are rather clear on this point: no-one "owns" space, even less celestial objects. And while the notion of "space sanctuary" has been long discarded from the American national rhetoric (as shown in Chapter 6), there is a consensus over the concept of space as a common.

A way to conciliate these two apparently contrasting aspects is what has been previewed by the US legislators, quickly followed by Luxembourg, in the sense of claims on the resources once mined, but not of the mining site itself. Not every country has equipped itself with a regulatory framework for such activities to take place: as a matter of fact, only a minority has any provision at all at a national level for what concerns the space sector as a whole, as shown in Chapter 7. It is worth observing that these attempts to create a domestic legal framework of space laws—clearly aimed at a civilian use of space—have stemmed from initiatives in the mining sector, and not from, for example, the space tourism, even though it is true that any kind of activities in this sense requires as a prerequisite the existence of a solid legal framework to develop. Lacking this, space tourism, however, conceived will probably remain a billionaires' fancy and nothing more.

We are still at an early stage. The missions aimed at returning samples from asteroids have only been (to date) a selected few, after that Hayabusa 1[15] became in 2006 the first spacecraft to land and grab fragments from a NEO and managed back to Earth in June 2010. Some of them are on their way—such as Hayabusa 2[16] and OSIRIS-REx—and, once they're back in 2020 and 2023, the science they're going to deliver will be key in terms of the future of space mining.

8.3 The Future of Space Tourism

When on 13 December 2018[17] Virgin Galactic flew up to 51.4 miles before returning to Earth, the international press was quick to label the company's success as "the boldest step yet in space tourism" (Phys.org 2018, online).

The claim is not unsubstantiated, even though more it is needed to start up the sector in earnest and take it further than what it is at the moment: the expensive hobby of a few billionaires.

The concept itself is quite recent; it was only in 2001 that the first space tourist ever, the American Dennis Tito, went to visit the ISS[18] on board of the Russian Soyuz, the same spacecraft that takes the astronauts to the station. This was the pattern for the following visitors, Mark Shuttleworth and Greg Olsen in 2005, and Charles Simonyi in 2007. The company that organised the trip was Space Adventures,[19] which negotiated a deal with the Russian agency Roscosmos for a seat on the Soyuz. In between 2001 and 2009, Space Adventure/Soyuz made the experience possible to seven other people,[20] for a price that was in the region of US$20–35 million. Otherwise said, about the half of NASA pays Soyuz for taking the US astronauts to the ISS.

If the concept of space tourism is these days' novelty, sending people into space is not, albeit it is still fairly restricted in terms of number and nationalities: since Gagarin came back from space in 1969—almost fifty years ago, less than 1000 people have so far left the Earth's surface if we consider long-haul flights still on planet.

As of December 2018, less than 600 people[21] have gone to space; they came from all over the world (38 countries), but roughly two-thirds were from the USA.

It is interesting to notice that this already limited number can become even smaller depending on which definition "space" one decides to accept. This is not merely a detail: the development of the sector will also mean a regulatory framework of reference, which will define where space is supposed to start—and where the specific regulations will apply.

According to Federal Aviation Administration (FAA)[22] guidelines, space is everything equal or beyond the so-called Kármán Line, which stands at 100 km/62 miles above sea level. FAA defines a suborbital spaceflight as a spacecraft that "reaches space but its velocity is such that it cannot achieve orbit" (FAA 2018, online). However, the US Department of Defence's classification is more inclusive and awards astronaut wings to all the pilots who fly higher than 80 km (50 miles).[23]

A more precise figure is available for the cosmonauts that have left the Earth's orbit.[24] So far, only 24 Americans travelled beyond Lower Earth Orbit (LEO) to go to the Moon; half of them walked on the lunar soil in six successful landings (Apollo 13 and Apollo 18 had a different story, briefly recap in Chapter 9). Since 1972, no human has left the Earth's orbit.

Who were all these few hundred spacenauts?

Most of them were proper astronauts—therefore, people with at least two years of training and often a PhD in astrophysics and mechanical engineering—but there have been in time more composite crews, in particular between 2001 and 2009. There were a few ISS-bound scientists, engineers, a half dozen or so Air Force pilots, and, as mentioned above, a few space tourists. And here a new set of issues, challenges and opportunities arose, together with legal implications and insurance requisites, which will all need to be addressed and solved if a second phase of space tourism is to begin.

The first important distinction is the typology of flights a company is going to offer. Different kinds of travels have to be labelled accordingly,[25] depending on how many kilometres above the sea level the plane is going to fly. For the moment, the only distinction is between orbital and suborbital flight. In the realm of this second category, two companies among all the others have captured the world's attention: Branson's Virgin Galactic (founded in 2004) and Bezos's Blue Origin (that started its operations as early as 2000). Not only they are racing against time to offer a seat to a wider range of consumers compared to the first space tourists; they also have an entirely different approach to the concept itself.

The originality of Virgin Galactic is in the design: the company's aim is to develop an enhanced "spaceplane" that can reach the limit of space, on a suborbital flight, at a comparatively affordable cost, while all the others, Elon Musk's SpaceX and Jeff Bezos's Blue Origin among them, use instead vertical rockets.

Branson's company was not the first to test this concept; the first suborbital flight of a rocket-powered spaceplane was SpaceShipOne, developed by Scaled Composites, a Californian-based subsidiary of Northrop Grumman. SpaceShipOne won the US$10 million Ansari X Prize for its successful flight attempt on 4 October 2004, becoming the first private crewed spacecraft to reach the boundary of space. The concept was so appealing that Virgin Galactic actually used a smaller model, a Scaled Composites SpaceShipTwo, VSS Enterprise, for its testing, including the one that ended in a spectacular crash in 2014[26] (Space.com 2014).

Since then, Virgin Galactic has adopted another model, SpaceShipTwo, VSS Unity, which has been developed instead by the Anglo-American Spaceship Company.[27] If anything goes according to plan, VSS Unity will launch together with its carrier aircraft WhiteKnightTwo and six passengers onboard and, once reached the altitude of 50,000 feet (about 15,000 metres), Unity will power up its own hybrid rocket engines and continue up to the limit of space (Space.com 2018a). The company is due to start in 2019 (Phys.org 2018), with Branson among the first passengers. According to the press, about 650 people are in the waiting list, having secured their seat for US$250,000.

Blue Origin, on the other hand, has followed SpaceX in the effort of developing a more traditional way to reach space, with a rocket called New Shepard.[28] A tiny capsule, fixed on the top of the 18-mt rocket—just as in the Mercury and Apollo days—will detach and carry six passengers to the same altitude of VSS Unity for then descending down to Earth.[29]

As it stands, the company has announced the start of its commercial operation at some moment in 2019.

Virgin Galactic and Blue Origin were not the only private companies that invested in the development of suborbital flight tourism. As it happens, the field is constellated by failures and bankruptcies of their competitors, another proof, if needed, of the challenges the sector presents. A well-known case is XCOR Aerospace,[30] a company that was developing a suborbital vehicle called Lynx, an independent spaceplane smaller than Spaceship 2 that could reach space without a carrier. XCOR announced the halt of Lynx development in 2016 and filed for bankruptcy in 2017. Another example is the European EADS Astrium, of the Airbus group, of a concept closer to Virgin Galactic than to Blue Origin, and whose plans for space tourism, originally set for 2012 (BBC News 2007), were later put on an indefinite hold.

Orbital tourism is a different story, mainly because the private companies here are the same already discussed in Chapter 3, and therefore, the ones already launching rockets for commercial flights. SpaceX and Boeing are the obvious providers in this field, since they are both well advanced in their development and have recently obtained their accreditation with NASA to transport astronauts to the ISS (see Chapter 3). At some moment in the early 2020s, they will also offer private passenger flights as a natural step forward. On the same way, Soyuz has announced its intentions to take tourists back to the ISS (Phys.org 2018).

A newcomer in this subsegment is the Houston-based company Axiom Space, which is currently selling a 10-day orbital trip package for US$55 million. The package will include transportation and stay on the ISS and a 15-week astronaut-training programme.

Axiom Space is a company to be watched closely for another reason[31]: as the already-mentioned Bigelow Aerospace, it is on its way to set up its own orbital station. Advertised on the company's website as "The World's First Commercial Space Station", Axiom will work both as orbital hotels (accommodate paying passengers by 2022) and manufacturing centres. In the same way of Bigelow's inflatable modules, Axiom will be linked to the ISS and detach once the construction and setting up is completed (*when* it is still uncertain; Space.com 2018b).

What is clear is that, once the transportation means will be in place to take tourists into the Earth's orbit, the number of the companies developing and offering facilities to welcome them onboard will increase. This is also the bet of another start-up, Orion Span, and its project of an orbital luxury hotel called "Aurora Station". The hotel should be ready in 2021 and open its hatches to visiting guests in 2022.

Space hotels have also been proposed for the Moon,[32] mainly as cislunar orbital stations[33] (Orbital ATK), and this is another subsegment that will probably see a development in the 2020s. Space Adventures is in the process of developing tourist rides that go to the Moon on a circumlunar trajectory. The most likely to succeed into taking tourists to the Moon is, again, SpaceX, whose owner Elon Musk has already declared the intention of sending two private passengers to the Moon on a circumlunar mission, following the flight plan of Apollo 8. The company plans to use BFR (Big Falcon Rocket) currently in development, for the mission (The Verge 2018).

Another point that has not been much discussed, but that is becoming increasingly popular due to the availability of suitable technology, is virtual space tourism. There are many opportunities now to experience the virtual reality of space, which can be quite an experience, both with real images taken from other celestial bodies, others with simulation, as in NASA's Exoplanet Travel Bureau that let the user choosing a destination about a series of exoplanets studied by the agency.[34]

Two UK-based companies, the VR-studio Rewind and the space expert of In-Space Missions, recently created SpaceTime Enterprises.[35] The company plans the launch of multiple satellites that will broadcast videos of the entire Earth and offer them to users in 3D virtual reality settings. The first satellite is scheduled to be launched in September 2019. Among other firms that are working in the same direction, there is the British Earth-I[36] and the American SpaceVR.[37] More could be done. For example, a rover could be sent on the Moon and simulate a "real driving experience" for the customers on Earth.

A real challenge in the steady development of space tourism might surprisingly be not technical. A few scholars (Apel 1997; Forganni 2017) have already highlighted the lack of a dedicated legal framework for space tourism, in a parallel to a lack of regulations for private endeavours in space. Until these aspects have been not properly sorted out, it is unlikely that space tourism will gain momentum, if anything for the enhanced level of risk involved (the already-mentioned Virgin Galactic's crash of 2014, with one casualty involved, is just a reminder, if any were indeed needed). While outer space tourism will always involve a higher level of danger compared to others (in the same way extreme sports like climbing the Himalayas or deep diving carry), a set of safety regulations are essential to make these risks as reduced and controlled as possible.

There are also other issues, both ethical and environmental to be considered, as the questionable (or not) use of space for entertainment purposes—1968 Outer Space Treaty (OST) rules out any appropriation of outer space—and the possible adverse effect on the environment, which space tourism can make it together better—by lifting pressure on some hard-impacted tourist area on planet—or worse—by affecting outer space, an area off-limit until now.

For all these reasons, an international legal framework which compounds and integrates domestic regulations on space tourism will prove essential to ensure the sector's development. A recent report (Digital Journal 2018) gives for the sector a global market size of US$490 million in 2017, with a perspective of growth up to US$1640 million by the end of 2025.

8.4 The Next Satellite Generation: CubeSat

If commercial satellites represent the lion's share of the global space industry from at least the 1990s, the new century has witnessed the rise of a new category of satellites: the so-called nanosatellites,[38] among which CubeSats are the most famous.

After having populated LOE for about fifteen years, they are now considered for interplanetary missions; in the last—to date—NASA's mission to Mars, InSight, two CubeSats[39] were the heroes of the day in the occasion of the probe landing on the Martian soil[40] on 26 November 2016, providing the relay with the Earth for telecommunication.

CubeSats have first made their appearance in 1999, as the result of cooperation between Jordi Puig-Suari, a professor at California Polytechnic State University (Cal Poly), and Bob Twiggs, a professor at Stanford University's Space Systems Development Laboratory (SSDL).

The CubeSat has certainly achieved its original intent, as NASA declares, which was to provide affordable access to space for the university science community. In fact, "thanks to CubeSats, many major universities now have a space program. But it's not just big universities; smaller universities, high schools, middle schools, and elementary schools have also been able to start CubeSat programs of their own" (NASA 2017: 5). The first six CubeSats were launched in June 2003 from Russia's Plesetsk launch site and soon made proselytes.

In technical terms, CubeSats belong to the category of small satellite—that is, any satellite less than 500 kg—but, different from others; they have to conform to certain specs to qualify as such. The reason behind the standardisation is, of course, the cost, as it makes possible for suppliers to manufacture component parts for a later assembling. This standardisation not only allows savings in manufacturing, but also in ground transportation and during the launching phase.

Contrary to what is generally believed, CubeSats is not a single measure satellite, but comes into different configurations all based on the CubeSat "unit"—the 1U. The 1U CubeSat is a 10-cm cube and a mass of 1 (max 1.33) kg. This was the original size, which, in time, has evolved in larger versions—1.5U, 2U, 3U, up to 6U. Now the CubeSats come as a preconfigured, ready-to-be assembled satellite "kit" which is especially suited for science experiment and observation.

Another important component of the CubeSat concept is the so-called dispenser, e.g., the interface between the satellites and the rockets that launch them, and this is where the standardisation helps, too. The CubeSats all fit well into the dispensers designed to house them for launch.

The first, and still operational in its latest revisions, is the Poly-Picosatellite Orbital Deployer (P-POD). It was developed by Cal Poly, which was designed to host any combinations of satellites up to 3U, together or in separate pieces. In NASA's words, "when it's time to release the payload, the launch vehicle sends an electrical signal to the P-POD, which causes the door to open and release the CubeSat(s) into orbit. Although most dispensers on the market have different designs, they all tend to follow the same basic idea of a safe container with a door that opens at the launch vehicle's command, which then ejects the CubeSat into space" (NASA 2017: 21).

In times, other dispensers[41] have developed, and now the offer varies.

In terms of rockets, things are somehow different, as virtually any rocket can take the CubeSats into space. As of 2017, some specific rockets have been used to carry them, namely Super Strypi (a small 17-mt rocket), Minotaur I Minotaur IV, Taurus XL, Delta II, Antares Falcon 9 and Atlas V. Others (Electron, Falcon 9 Heavy, LauncherOne, Space Launch System—SLS) have put CubeSats in their logs of upcoming flights.

They are deemed to grow in number, for a trend that has been already noticed in the launching industry, i.e., the worldwide development of small launching vehicles aimed at catering this specific demand. There are at the moment at least 33 small ($\leq$500 kg) launch vehicles[42] under development, specifically aimed at cater small payloads into LOE at competitive prices (SIA 2017), even though the uncertain funding situation makes their feasibility unknown. NASA holds a CubeSat Launch Initiative[43] aimed at finding free launching slots for CubeSats as a piggy-back of already scheduled rocket launches, and so does ESA, with the initiative "Fly Your Satellite!",[44] using its light Vega launcher.

CubeSats are not the only existing model of nanosatellites. There are others[45], even less expensive to put into orbit.

The current cost of sending CubeSat into LOE (this is the working altitude for all the small satellites) is about US$40,000. However, TubeSat,[46] another nanosatellite model developed by InterOrbital Systems (IOS), reduced that entry-level price down to US$8000, including the satellite itself. Like CubeSat, TubeSat can be assembled into 1, 2 or more units, and it looks at the academic and/or educational institutions[47] at its first market.

There are two main hindrances to the nanosatellite models: first of all, the size of their electronics (their minuscule payload imposes constraints), which makes them more sensitive to radiation, sun weather, and cosmic rays. Moreover, they are put into a self-decaying LOE (to avoid increasing the already-high number of debris), which means that their average life is only of a few months at best, before falling back and incinerate in the Earth's atmosphere.

Albeit this structural weakness, the small satellite numbers keep growing, and the global statistics[48] for nanosatellites are nothing but impressive (Fig. 8.1).

As of October 2018, there were 966 nanosats launched into space, 878 of which CubeSat. 86 of them were destroyed at launch. 323 companies have produced nanosatellites of some kind, for a total of 58 countries. The record level of a number of satellites on the same launch is 103 (by India, in February 2017). The forecast is of over 3000 units to be launched in six years' time (Fig. 8.2).

Not all the missions are the same in terms of choices of size.

A closer look would show that the 1U size is the preferred one by the universities, while industry-led missions select 3U or bigger. This does not come as a surprise. As a matter of fact, the distinction between the two "also underscores the different cost structures between industry and

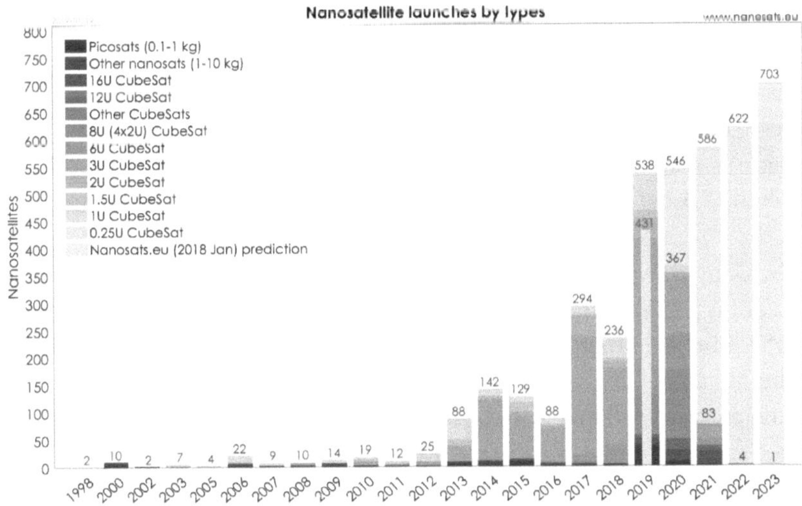

Fig. 8.1 Cubesat statistics, numbers (*Source* Erik Kulu, Nanosatellite and CubeSat Database, www.nanosats.eu)

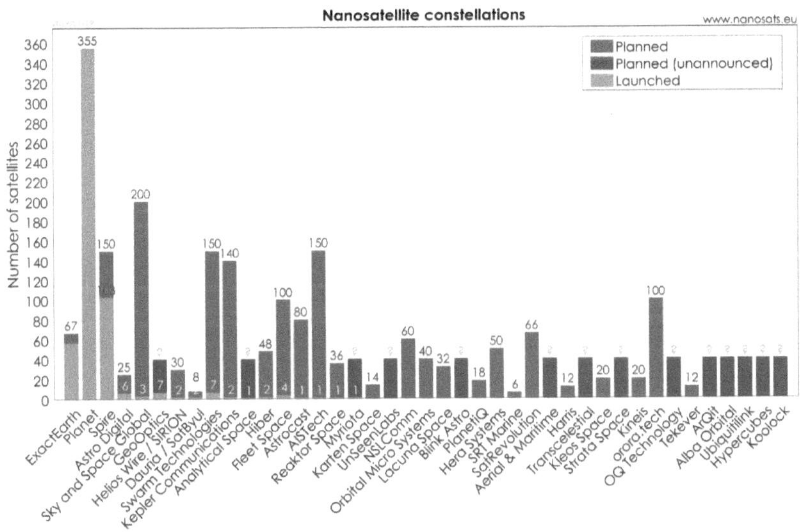

Fig. 8.2 Cubesat statistics, typology (*Source* Erik Kulu, Nanosatellite and CubeSat Database, www.nanosats.eu)

university missions: for a professional mission, the costs of launch and of spacecraft hardware are small compared to personnel expenses, whereas for a university, the cost of the launch and spacecraft hardware are the largest components of the mission cost. Thus, the cost of a professional mission is similar whether the spacecraft is 1U or 3U, while a university 3U could cost three times as much as a university 1U" (Swartwout 2013: 15).

Breaking down further the numbers, one segment where the CubeSat's presence is particularly prominent is in the Earth Observation subsegment. Of 55 CubeSat launched in 2016, 45 were for EO (and most of them being of the 3U size).

The company Planet Labs sent 32 of them, and it maintains dozens of CubeSat-sized satellites in orbit, using them in disaster management to climate monitoring.

The CubeSats have not only changed the perception and the concept of satellites and made them available to a category of users, such as research centres, universities, organisations, which were until that moment outside the space sector Menaa (2016); they have also started to influence the way the operators of commercial satellites do business as well.

Even big constellations of commercial satellites have started using customised small units, albeit larger than the typical CubeSats; companies such as Airbus/SSTL, SSL, Spaceflight Industries, York Space Systems lead the way.

The miniaturisation does not stop at satellites. Among the most daring ideas of the new century, there is the Breakthrough Initiative, with the idea of nanocrafts for interstellar exploration, which has been briefly mentioned in Chapter 9.

Case Study: A Space for Everybody—ESA Rosetta

There have been a few moments in the last sixty years when space was on everybody's mind and discourse, and many still remember the excitement of the Moon landing and the Apollo programme. While those days have gone and the adventure of space has known a long period of quiet—if not lack of interest—there is now a renovated attention to the sector, with a substantial difference compared to the past: engagement, and the increasing involvement of the civil society.

"2018 will also see space become accessible to more people. This will be in line with increased venture capital investment in start-up space companies and the growth of boutique space agencies focused on niche areas. There is a growing number of Australian small-to-medium businesses seeing challenges and opportunities presented by this new era in space –Space 2.0" (CsiroScope 2018, online).

What is true for Australia—which just established its own space agency in 2017—it is true for the rest of the world, too. There are so many initiatives, some of which have been ongoing for long and that have found a new impetus, aiming at engaging people into the space adventure.

This case study mentions a few of them; far more is out there. This is just a starting point.

The World Space Week[49]

Since its United Nations declaration in 1999,[50] World Space Week has grown into the largest public space event on Earth. Celebrated on October 4–10 every year, it has seen a growing number of globally held activities and events. "World Space Week consists of a myriad of space-related events held by space agencies, aerospace companies, schools, planetaria, museums, and astronomy clubs in a common timeframe to achieve greater student and public impact through synchronisation" (UN-World Space Week 2018, online).

In 2017, there were more than 3700 events in 80 countries celebrating the benefits of space exploration. The 2018 theme was "Space Unites The World" and it was inspired by UNISPACE+50, a historic gathering of world space leaders that occurred in 2018 with the aims of promoting cooperation between spacefaring and emerging space nations.

The theme of the World Space Week 2019 is instead "The Moon: Gateway to the Stars", as announced by World Space Week Association (WSWA), to celebrate the 50th anniversary of humankind's first footprints on the Moon.

Yuri Night

Another global event is represented by the Yuri Night, which in turns celebrates the Soviet cosmonaut Yuri Gagarin in his historic voyage on the Earth's orbit on 12 April 1961 since 2001. Everybody can go to the website and register their own events and celebrations. To give an idea of the size, in 2015 Yuri Night[51] was celebrated in 7 continents, 49 countries, with a total of 223 events.

Other initiatives focus instead on making possible the direct involvement of people in space activities and research.

Zooniverse

Zooniverse, otherwise called "people-powered research", is a global platform where volunteers enrol and help professional researchers on their investigation. The platform is not exclusive for space studies; it is a well-known citizen science project, where volunteers work on data professionally collected to help analyse them in many fields, such as Earth Science, Zoology, and other disciplines.

A quick look at the site Zooniverse will also show, however, that many projects are actually devoted to astrophysics and space in general—and, with more than 1 million people (the number was reached as early as 2014), the platform has made valuable contributions to space sciences (more than 55 articles in peer-reviewed scientific publications have been recorded in 2014; Smith et al. 2011).

In 2011, initiatives like Galaxy Zoo and Planet Hunters have helped professional astronomers discover new exoplanet using Kepler data, while other studies aimed at the identification of asteroids (Schwamb et al. 2010; Shannon 2014). One of the ongoing space projects is Planet Hunter Tess Statistics, which works on the huge amount of data provided by the recently launched Transiting Exoplanet Survey Satellite (TESS)—Kepler's successor—to look for planets outside of our own Solar System, including planets that could support life. So far,[52] the project has enrolled 4775 volunteers and completed 126,925 analyses.

Ultrascope

Ultrascope is a robotically controlled, 3-D printed, laser-cut, "ultra low-cost" telescope, which was originally designed during the 2015 International NASA Space Apps Challenge by London-based Open Space Agency (OSA). The concept behind it is that anybody can download its files and build its own professional telescope at home. In its original version, the Ultrascope Explorer (or UEX) used a high megapixel smartphone as the camera and input/output device (i/o) using 4G LTE. Based on this design, the students of a small town in South African connected in 2015 with Wired Business Conference in NYC and broadcast live images of the African sky.

Asteroid Grand Challenge (AGC) Initiatives

Another example of citizen science is NASA's AGM, whose many initiatives were often made possible by the high level of involvement of the civil society.

"On June 18, 2013, NASA announced the Asteroid Grand Challenge to find all asteroid threats to human population and know what to do about them. Since then, NASA has inspired asteroid hunters, met citizen scientists from across the globe, and challenged people to ask themselves if we are smarter than the dinosaurs. NASA is creating new ways every day to get the global public more involved in protecting the planet from asteroid threats" (NASA-AGM 2014, online).

Especially the project Asteroid Data Hunter received a great degree of attention by the public, which resulted in a record number of registrations and solutions. Students have been engaged in the asteroid detection using, as in the case of Zooniverse, data collected by professional asteroid survey teams. In the first pilot initiative that involved 7th Graders from Dillard Drive School in Raleigh, North Carolina, the students discovered 10 new asteroids.

The AGC concept attracted a lot of attention, serving "as a beacon for high-impact, multi-disciplinary collaborations among the government, industry companies, universities, non-profits, philanthropists, and individuals including scientists and engineers. A number of new formal and informal partnerships emerged as a result of the AGC that may not have been pursued otherwise: for example, the SAA with SpaceGAMBIT and Verizon as well as the ongoing engagement of Intel and Nvidia with the FDL" (Gustetic 2018: 10).

There is no doubt that nobody more than NASA contributed to popularise space and promote engagement by the civil society since the 1960s (Apollo Programme was on everybody's mouth when the USA landed on the Moon).

While the degree of interest and engagement ebbed away in the 1970s and following years, to come back in the new century until today's new "boom", the agency has continuously worked to reach out to the big public. Recent missions, as New Horizons and InSight (Mars lander), have attracted a lot of attention, communicated as they are to the public in a clear and relatively untechnical vocabulary, and with the clear intention to promote engagement.

A quick look at the NASA's website will discover a trove of projects and activities even for young kids, together with projects of citizen sciences of the kind of Zooniverse. One now ongoing is GLOBE Observer,[53] where users can participate to create a more detailed satellite-based global maps of the Earth.

However, another agency has recently achieved something at least as important: an inspirational mission that polarised the world as much as in the early days of the Apollo programme, even without involving humans: ESA, with its Rosetta mission.

ESA & Rosetta

Rosetta has become a textbook case in the sense of a mission that managed to keep millions of people glued to their screen for hours during the key phases.

Launched on 2 March 2004, the spacecraft Rosetta, together with its lander probe Philae, achieved the historical result of orbiting and then landing on the surface of a comet—67P/Churyumov–Gerasimenko in August 2014. After two years of detailed study of the comet, on 30 September 2016, Rosetta ended the mission by crashing on the comet's surface. What was unique was ESA's media campaign: the agency requested the creation of an illustrated fairy-tale cartoon describing in a series of videos the story of Rosetta using the "Sleeping Beauty" theme, in which both Rosetta and Philae were given anthropomorphic personalities.

The series of videos, called "Once upon a time…", was made available on the internet in different languages and displayed the phases of the missions in a language easy to understand even for young students.

"The spacecraft cartoon characters also gained popularity in synergy with the Twitter accounts @ESARosetta (managed by ESA's science communications team) and @Philae2014 (managed by the German Aerospace Centre, DLR), which were run in first person, letting the public engage with the characters of Rosetta and Philae on their adventure to explore Comet 67P/Churyumov-Gerasimenko" (Marcu and Laird 2016, online).

Once the mission was over, the Twitter accounts continued to be maintained (still are), twitting with regularity in some funny and easy-to-read comments over the news, and actively engaging with space enthusiasts. This might explain why Rosetta and Philae are still so popular especially among school kids.

ESA's Chief Strategy Officer Kai-Uwe Schrogl (2017) commented on the exceptional value of ESA Rosetta's mission: "Rosetta and Philae inspired. They accomplished a first and they did so unimaginably far away from Earth in breath taking reality. So, inspiration as a means for as well as a result of popularisation is unbeatable. It always is authentic and it makes you wonder what can be accomplished now and what we would like to contribute to accomplishments in the future. We have to make efforts to prepare new narratives that actually touch people, making space tangible and helping to reach the level of human empathy" (Schrogl 2017: 71).

While Rosetta remains ESA's most successful initiative in terms of getting space near to the public and popular again, it has not been the only one. There have been a few initiatives over the year, from surveys to widespread debates meant to gather the public interest in the subject.

As Schrogl himself acknowledged, "popularisation can work through surveys. They meet the double objective of learning what people think and at the same time they engage people. Space in Europe has continuously been a topic for surveys mostly run by Eurobarometer on behalf of the European Commission. They lead to good information and intelligence. Sometimes, however, the client is happy about the result that almost 100% are in favour of using Earth observation to monitor the climate, but this delight is immediately reduced, when in the next question zero percent can name a single EO satellite mission" (Schrogl 2017: 72).

One, organised by the agency in September 2016, was a "Citizens' Debate on Space", in which about 100 persons met for a day in each one of ESA's 22 Member States in moderator-led sessions, covering all the major aspect of space-exploration in a more focused, Delphi-based, opinion-gathering exercise which addressed the common shortcoming of surveys.

That ESA has managed well in making its activities known to the big public has been proved by a recent survey just released by the space agency. Carried out on behalf of ESA by Harris Interactive in December 2018, the survey asked 5000 people aged 18 or older from Europe's five most populous countries (Germany, UK, France, Spain and Italy) a few questions about space. According to them, there are three main reasons to invest in space, namely, a better understanding of the Universe, a better knowledge of the Earth, and to help make life on Earth easier. About 83% have heard of ESA before—and this is one of the things Rosetta has certainly contributed to—even though not everybody knew the details[54] of ongoing space projects.

The most important result was probably the general attitude towards space, about which about 90% of respondents had a positive view: something to build on for the future.

Sources

CsiroScope. (2018, January 29). *Australia Takes Up Space in 2018*. Available at https://blog.csiro.au/australia-takes-up-space-in-2018/.

Marcu, S. D., & Laird, R. J. M. (2016, March 13). Captivating Hearts and Minds. *Room, The Space Journal*. Aerospace International Research Center. Retrieved 28 December 2016.

Space Daily. (2018, September 14). *Help Make a Better World Land Map with NASA App*. http://www.spacedaily.com/reports/Help_make_a_better_world_land_map_with_NASA_App_999.html.

Zooniverse. (2017). *List of Publications*. Available at https://www.zooniverse.org/about/publications. Retrieved on 11 February 2018.

Notes

1. Good enough to discover Jupiter's four moons, which have therefore being called "Galilean Moons".
2. https://www.space.com/14075-10-biggest-telescopes-earth-comparison.html10. Biggest Telescopes on Earth: How They Measure Up, 29 December 2011, 01:22 p.m. ET.

3. "ALMA studies the building blocks of stars, planetary systems, galaxies and life itself. By providing scientists with detailed images of stars and planets being born in gas clouds near our Solar System, and detecting distant galaxies forming at the edge of the observable Universe, which we see as they were roughly ten billion years ago, it lets astronomers address some of the deepest questions of our cosmic origins." ESO-ALMA, In search of our cosmic origins. https://www.eso.org/public/usa/teles-instr/alma/.
4. Among Hubble's achievements, the detection of early-Universe galaxies only after 600 million years after the Big Bang detected in 2010 in the Hubble Ultra Deep Field and, in 2008, one of the first direct snapshots of an exoplanet, Fomalhaut b, orbiting its parent star. Hubble got recently in trouble in October 2018 after one of the three gyroscopes (gyros) failed. The telescope was put into safe mode while NASA worked to solve its issues and resumed services a few weeks after. http://hubblesite.org/.
5. Chandra X-Ray observatory. http://chandra.harvard.edu/about/.
6. 2018 has proved to be a challenging year for space telescope. In just a couple of months, a series of failures where registers on board of Chandra, Hubble, and Kepler ran out of its operation life. Albeit their services have become reliable and regular, the hurdles to maintain them operational can't be discounted, both in terms of costs and efforts, enhancing the needs for alternative solutions.

   Behind the Scenes of Recovering NASA's Hubble 27 November 2018, 11:30 a.m. (EST). http://hubblesite.org/news_release/news/2018-61.
7. James Webb Space telescope, 2018. https://jwst.nasa.gov/index.html.
8. This book uses data from ESO's press release of January 2018. Other sources, such as space.com, reports instead a primary mirror of 138 feet (42 metres) ESO European Southern Observatory Official website: https://www.eso.org/public/.
9. When operational, TMT will provide new observational opportunities in essentially every field of astronomy and astrophysics. Observing in wavelengths ranging from the ultraviolet to the mid-infrared, this unique instrument will allow astronomers to address fundamental questions in astronomy ranging from understanding star and planet formation to unravelling the history of galaxies and the development of large-scale structure in the universe. https://www.tmt.org/page/about#what-is-tmt.
10. A good description is provided here. http://sci.esa.int/cosmic-vision/54517-athena/.
11. Gravitational waves were predicated by Einstein.

In October 2017, a pool of telescopes detected them for the first time, as a by-product of a merger of two binary stars.

12. "The final design that the students came up with for the AAReST spacecraft features a central core unit that contains the attitude and control system, the radio and the antennas, and two rigid mirrors. Surrounding that central unit are two smaller spacecraft containing deformable mirrors that will be co-focused with the rigid mirrors on the main spacecraft. The whole package is small, measuring just 46 x 34 x 30 centimetres. Once in orbit, the two smaller spacecraft will detach from the main one via splitting bolts and then float about 30 centimetres away from the spacecraft before reconnecting to the main body in a wider configuration via electromagnets" (Caltech University 2018, online).
13. Keck Institute for Space Studies (2012) "Asteroid Retrieval Feasibility Study." http://kiss.caltech.edu/final_reports/Asteroid_final_report.pdf.
14. "For 40 years astronauts have depended on Earth for resupply and operational support. Many missions aboard spacecraft like the Apollo capsules or space shuttles lasted only days or weeks. Typical stays aboard the International Space Station, in low-Earth orbit, are six months. Crews aboard the space station can return to Earth in a matter of hours in the case of an emergency. We call these missions "Earth Reliant. Testing aboard the space station is helping us develop ways to break these Earth-reliant bonds, so astronauts can be more autonomous the farther into the solar system they explore. The ARM robotic mission and crewed mission to explore the asteroid will further advance these capabilities in the 'Proving Ground' between Earth and Mars, or what we call cis-lunar space--the area around the moon" (NASA 2014, online).
    NASA (2014) How Will NASA's Asteroid Redirect Mission Help Humans Reach Mars? 27 June 2014. https://www.nasa.gov/content/how-will-nasas-asteroid-redirect-mission-help-humans-reach-mars.
15. Incidentally, the ongoing Japanese mission Hayabusa 1 visited (and collected samples) from one of them, the 300-m-diameter asteroid 25143 Itokawa.
16. Those samples are still under study. "Hayabusa2 aims to collect three samples from different locations on Ryugu, netting approximately 100 milligrams. OSIRIS-REx will attempt to collect up to two kilograms from a single spot on Bennu's surface" (Hsu 2018, online).
17. The flight also marked the Virgin Galactic's first mission for NASA, carrying on board a few science experiments in its payload.

18. Interestingly, Tito had originally bought the ticket for the Russian station MIR. But, as Mir was decommissioned in 2001, Space Adventures "rebooked" him on the ISS.
19. Space Adventures remains, almost 20 years later, the only company to have sent paying passengers to space, a fact that alone tells how challenging the whole concept can possibly be. And if, since 2009, it has been not possible to take visitors to the ISS any longer, the company has continued its long-term relationship with the Russian, promoting visits to the Soyuz launches and zero-g experiences (see the company's official website: http://www.spaceadventures.com/experiences/zero-gravity-flight/).
20. The last visitor was Guy Laliberté, the founder of Cirque du Soleil. These are considered the early days of the space tourism that came to an end in 2009 for a series of reasons, the doubling of the permanent crew on the ISS and the decommissioning of the shuttle in 2011, which left the Soyuz alone as a carrier for astronauts.
21. This relatively imprecise figure has been provided for a reason: there are different numbers, 561, 538, 533, from different sources, and in a way, they are all correct, because they refers to different ways to define what the limits of space are, as better specified in the text.
22. The US Federal Aviation Administration (FAA) is the entity responsible for commercial space launches, including licensing and authorisation.
23. According DoD, eight people qualified with a rocket-assisted X-50 spaceplane, but only one of them flied higher than 100. Hence, the difference in the statistics.

    Rocket-assisted spaceplane have not being the rule in suborbital flight (even though recently Scaled Composites SpaceShipOne have modernised the concept). Apart from them, all the other astronauts have instead gone suborbital in non-winged rockets or in shuttles.
24. Other statistics are also of interest: apart from the Apollo missions, the space shuttle programme was the ones carrying the bulk of astronauts. For reasons of space, there has been no section devoted to the space shuttle and its long-serving records. A few of them have been mentioned in the previous chapters when appropriate. It is only the case to highlight here that the space shuttle family, with its six spacecraft, is rightly famous, representing at the time (as already discussed) a revolutionary concept and proving essential for a whole series of projects, ISS and Hubble ST among them.

25. This is not just a matter of label. A clear definition is paramount because a different kind of legal framework—involving insurances, definition of liability and requirements—is applicable or not to high-altitude flight.
26. The crash was serious: it left one of the two pilots dead, the other injured, and completely destroyed the spaceplane.
27. The company was founded by Branson and Scaled Composite's designer Burt Rutan in 2005.
28. The name comes from the first American man in space, Alan Shepard.
29. If testing is of any guide to the experience, Virgin Galactic will allow for a longer permanence in space: 90 minutes (up to 2 hours) against Blue Origin's 10 minutes. Details at Phys.org 2018. The price should be on the same range, about US$200,000–300,000 for Blue Origin.
30. Similarly, another American start-up, Armadillo Aerospace, was put on "hibernation" in 2013 for the growing difficulties of the project. Read the whole story here: http://www.newspacejournal.com/2013/08/01/carmack-armadillo-aerospace-in-hibernation-mode/, 1 August 2013.
31. While Axiom has been incorporated only in 2015, its founder and CEO, Michael Suffredini, is an old hand in outer space, having managed NASA's ISS program for a decade. See the company profile here: https://axiomspace.com/.
32. As much as in the suborbital flight segment, also the Moon trips see their own "graveyard" of start-ups that didn't manage to continue on their projects. Golden Spike, active in between 2010 and 2013 and that planned to send roundtrip missions to the satellite, was one of them. See: http://nasawatch.com/archives/2015/10/golden-spike-an.html. Another, even more ambitious, was the Artemis Project, linked to the non-profit Artemis Society (http://www.asi.org/). The Artemis Project was a commercial venture that aimed to establish a permanent, self-supporting crewed lunar base. The initiative has (so far) not advanced farther than the design stage.
33. One is planned by Orbital ATK, which uses its Cygnus spacecraft to deliver unmanned cargo and supplies to the ISS. The company announced its plans in May 2017: "By 2021 we would launch an initial habitat module aboard NASA's heavy lift SLS rocket with a crewed Orion capsule. This mission would serve to prove out the Orion capsule and deploy the initial habitat in cislunar space. When Orion and the crew return to Earth, the module would remain, operating autonomously, acting as a platform for experiments and serving as a destination for future crewed

missions." Read the press release here: https://news.northropgrumman.com/news/features/cislunar-space-habitation-paving-the-way-to-mars.

34. On the landing page of the application, the user land on (supposed) rocky and deserted landscape of TRAPPIST-1d. On the screen, a menu lets the opportunity to choose another world to explore. https://exoplanets.nasa.gov/alien-worlds/exoplanet-travel-bureau/explore-trappist-1d/?travel_bureau=true.
35. More about the company is available here: http://spacetime.enterprises/.
36. The company also offer a sample set of data on its website: https://earthi.space/.
37. Company website: https://www.spacevr.co/#spacevr.
38. The term "nanosatellite" is used here for all satellites in between 100 g and 40 kg, even though the satellite classification would give: microsatellite: 10 to 100 kg; nanosatellite: 1 to 10 kg; pico satellites 100 g to 10 kg. But since CubeSats and the other comes in assembling units, this book followed the convention also adopted by nanosatellite database, and called all "nanosatellite".
39. Named Wall-E and Eve, out of the Pixar movie, and collectively known as MarCO (Mars Cube One), the two CubeSat followed the mission running on Solar power and relying information to NASA. For a technical description of the two, see: https://www.space.com/40457-insight-mars-lander-nasa-cubesat-hitchhikers.html.
40. Upon that success, they'll probably be included in the new mission now planned to Jupiter (Space.com 2018c).
41. In some cases, there was no dispenser altogether, like in the case of the Peruvian Chasqui 1. Sent to the ISS in a resupply mission in 2014, the CubeSat was released into space by a cosmonaut during a spacewalk.
42. Other sources list 29, together with providing their details. See: http://www.parabolicarc.com/2016/10/03/plethora-small-sat-launchers/.
43. https://www.nasa.gov/content/about-cubesat-launch-initiative. More than 60 units have been sent so far into space in nine years of project.
44. NASA has a lot of documentation in this sense. See, for instance: http://www.esa.int/Education/CubeSats_-_Fly_Your_Satellite. As of December 2018, ESA supported 20 student CubeSat teams from ESA Member and Cooperating States.
45. In addition to the ones mentioned here, there's PocketCube. http://mstl.atl.calpoly.edu/~bklofas/Presentations/DevelopersWorkshop2014/Twiggs_PocketQube.pdf. SunCubes and other non-standard pico satellites.

46. TubeSat, which comes with a tubular, hexagonal architecture, 12 cm in length and 4 cm in diameter and the basic hardware, is more cost-effective of CubeSat. However, since it is a proprietary design, the customers are obliged to stick to IOS, while CubeSat, being an open design, can be more easily reconfigured and fit with a larger array of suppliers.
47. As reported on IOS website, a Tancredo Sat-1, an IOS TubeSat kit assembled by students from the Tancredo Middle School in Ubatuba, Brazil, was launched to the ISS by a Japanese H2 rocket and deployed from the Kibo module in January 2017. See: http://www.interorbital.com/Tubesat.%20Kits.
48. All figures here come from the database maintained at https://www.nanosats.eu/ and from SIA (2017, 2018).
49. Details and history of the initiative are available on the official website: http://www.worldspaceweek.org/.
50. "The General Assembly declares 4 to 10 October World Space Week to celebrate each year at the international level the contributions of space science and technology to the betterment of the human condition" Declaration at the *UN General Assembly resolution*, 6 December 1999.
51. For the complete yearly statistics, see: https://yurisnight.net/data/.
52. These statistics have been retrieved from https://www.zooniverse.org/projects/nora-dot-eisner/planet-hunters-tess on 17 January 2019.
53. The way this is done by sharing photos of the user's location with the module "Land Cover: Adopt a Pixel", identify the type of land (trees, grass, etc.) and then match the single observations to satellite data. The 10 citizen scientists who map the most land in between September 22, Public Lands Day, and October 1, NASA's 60th anniversary, 2018, will be acknowledged on social media and receive a certificate of appreciation from GLOBE Observer.
54. Another interesting result was that without exceptions all of them overestimated the cost of space activities for the public finances of their countries. For precise data about the survey, see: http://www.esa.int/About_Us/Welcome_to_ESA/How_much_do_European_citizens_know_about_space.

References

Apel, U. (1997). Space Tourism—A Promising Future? *Space Policy, 13*(4), 279–284.

BBC News. (2007, June 13). *Science/Nature. Firm Rockets into Space Tourism* [Online]. News.bbc.co.uk. Available at http://news.bbc.co.uk/2/hi/science/nature/6749873.stm. Retrieved on 17 January 2019.

Blair, B., Diaz, J., & Duke, M. (2002). *The Case for Commercial Lunar Ice Mining: Final Report to the NASA Exploration Team.* Washington, DC: National Space Society. Available at http://www.nss.org/settlement/moon/library/2002-CaseForCommercialLunarIceMining.pdf. Retrieved on 18 December 2018.

Caltech University. (2018, June 8). *Student-Built Satellite Telescope Prepares for Space.* Press Release. Available at http://www.caltech.edu/news/student-built-satellite-telescope-prepares-space-83205.

Chandra X-Ray Observatory. (2018). http://chandra.harvard.edu/about/.

Digital Journal. (2018). *Space Tourism Market 2018 Global Growth, Opportunities and Industry Analysis Forecast to 2025* [Online]. Available at http://www.digitaljournal.com/pr/3972533#ixzz5ZZh9agMP. Retrieved on 17 January 2019.

Elvis, M. (2014). How Many Ore-Bearing Asteroids? *Planetary and Space Science, 91,* 20–26.

Elvis, M. (2017). What Can Space Resources Do for Astronomy and Planetary Science? *Space Policy, 37*(2), 65–76.

ESA. (2017). *Near-Earth Objects—NEO Segment.* Available at http://www.esa.int/Our_Activities/Operations/Space_Situational_Awareness/Near-Earth_Objects_-_NEO_Segment. Retrieved on 30 November 2018.

ESO-ALMA. (2019). *In Search of Our Cosmic Origins* [Online]. Available at https://www.eso.org/public/usa/teles-instr/alma/. Retrieved on 10 January 2019.

ESO European Southern Observatory Official. (2019). Official Website. Available at https://www.eso.org/public/. Retrieved on 10 January 2019.

FAA. (2018). *The Annual Compendium of Commercial Space Transportation: 2018.* Available at https://www.faa.gov/about/office_org/headquarters_offices/ast/media/2018_ast_compendium.pdf.

Forganni, A. (2017). The Potential of Space Tourism for Space Popularisation: An Opportunity for the EU Space Policy? *Space Policy, 41*(2017), 48–52.

Foust, J. (2017, June 14). NASA Closing Out Asteroid Redirect Mission. *SpaceNews.com.* Available at https://spacenews.com/nasa-closing-out-asteroid-redirect-mission/. Retrieved on 30 November 2018.

Gustetic, J. L., Friedensen, V., Kessler, J. L., Jackson, S., & Parre, J. (2018). *NASA's Asteroid Grand Challenge: Strategy, Results and Lessons Learned.* https://arxiv.org/abs/1803.04564.

Hsu, J. (2018, August 1). Collecting Space Rocks. *Scientific American, 319*(2), 12–14. Available at https://doi.org/10.1038/scientificamerican0818-12.

Hubble Space Telescope. (2019). Official Website. Available at http://hubblesite.org/. Retrieved on 10 January 2019.

Keck Institute for Space Studies. (2012). *Asteroid Retrieval Feasibility Study* [Online]. Available at http://kiss.caltech.edu/final_reports/Asteroid_final_report.pdf. Retrieved on 10 January 2019.

Mainzer, A., et al. (2011). NEOWISE Observations of Near-Earth Objects: Preliminary Results. *The Astrophysical Journal, 743*, 156. https://doi.org/10.1088/0004-637X/743/2/156.

Menaa, F. P. (2016). New Opportunities Offered by Cubesats for Space Research in Latin America: The SUCHAI Project Case. *Advances in Space Research, 58*(10), 2134–2147.

NASA. (2014). *Asteroids Provide Sustainable Resource.* Available at https://www.nasa.gov/mission_pages/asteroids/news/asteroidmining.html. Retrieved on 1 February 2017.

NASA. (2015). *A Grand Challenge: Find and Plan for All Asteroid Threats.* Washington, DC: National Aeronautics and Space Administration; c2021–17. Available at http://www.nasa.gov/sites/default/files/files/FS-AGC_20130925_Tagged.pdf. Retrieved on 10 February 2018.

NASA. (2017). *CubeSat 101 Basic Concepts and Processes for First-Time CubeSat Developers.* Available at https://www.nasa.gov/sites/default/files/atoms/files/nasa_csli_cubesat_101_508.pdf.

NASA-AGM. (2014). *Be an Asteroid Hunter in NASA's First Asteroid Grand Challenge Contest Series.* Available at https://www.nasa.gov/content/be-an-asteroid-hunter-in-nasas-first-asteroid-grand-challenge-contest-series.

NASA-ARM. (2018). *What Is NASA's Asteroid Redirect Mission?* [Online]. Available at https://www.nasa.gov/content/what-is-nasa-s-asteroid-redirect-mission. Retrieved on 10 January 2019.

Phys.org. (2018, July 13). *First Space Tourist Flights Could Come in 2019.* Available at https://phys.org/news/2018-07-space-tourist-flights.html. Retrieved on 17 January 2019.

Popular Science. (2019). How to Become a Space Tourist: 8 Companies (Almost) Ready to Launch. Available at https://www.popsci.com/how-to-become-a-space-tourist#page-3. Retrieved on 17 January 2019.

Schrogl, K. U. (2017). The Popularisation of Space—A European Perspective. *Space Policy, 41,* 70–72.

Schwamb, M. E., Brown, M. E., Rabinowitz, D. L., & Ragozzine, D. (2010). Properties of the Distant Kuiper Belt: Results from the Palomar Distant Solar System Survey. *The Astrophysical Journal, 720*(2), 1691–1707. https://doi.org/10.1088/0004-637x/720/2/1691.

Shannon, H. (2014, January 20). *Zooniverse Reaches One Million Volunteers.* Universe Today. Available at https://www.universetoday.com/109413/zooniverse-reaches-one-million-volunteers. Retrieved on 18 December 2018.

SIA. (2017). *Satellite Industry Association's 2016 Report.* Available at https://www.sia.org/annual-state-of-the-satellite-industry-reports/2017-sia-state-of-satellite-industry-report/. Retrieved on 27 February 2019.

SIA. (2018). *Satellite Industry Association's 2017 Report.* Available at https://www.sia.org/2018_ssir/. Retrieved on 27 February 2019.

Smith, A. M., Lynn, S., Sullivan, M., Lintott, C. J., Nugent, P. E., Botyanszki, J., et al. (2011). Galaxy Zoo Supernovae. *Monthly Notices of the Royal Astronomical Society, 412*(2), 1309–1319.

South China Morning Post. (2018, July 13). Space Tourism with Virgin and Blue Origin, Months Away Yet Worlds Apart. Available at https://www.scmp.com/news/world/article/2155138/space-tourism-flights-virgin-and-blue-origin-are-mere-months-lift-their. Retrieved on 17 January 2019.

Space.com. (2011, December 29). *10 Biggest Telescopes on Earth: How They Measure Up.* Available at https://www.space.com/14075-10-biggest-telescopes-earth-comparison.html. Retrieved on 1 February 2017.

Space.com. (2012a). *Asteroid Mining Venture Backed by Google Execs, James Cameron Unveiled.* Available at https://www.space.com/15395-asteroid-mining-planetary-resources.html. Retrieved on 30 November 2018.

Space.com. (2012b, April 24). *Is Asteroid Mining Possible? Study Says Yes, for $2.6 Billion.* Available at https://www.space.com/15405-asteroid-mining-feasibility-study.html. Retrieved on 29 November 2018.

Space.com. (2014, November 3). *Virgin Galactic SpaceShipTwo Crash Investigation in Photos.* Available at https://www.space.com/27636-virgin-galactic-spaceshiptwo-crash-investigation-photos.html. Retrieved on 17 January 2019.

Space.com. (2018a, July 12). *Scaled Composites: Builder of SpaceShipTwo.* Available at https://www.space.com/19405-scaled-composites.html. Retrieved on 17 January 2019.

Space.com. (2018b, June 14). *Want to Take a 10-Day Trip to the Space Station? It'll Cost You $55 Million.* Available at https://www.space.com/40893-axiom-private-space-station-trip-55-million.html. Retrieved on 17 January 2019.

Space.com. (2018c, April 5). *'Luxury Space Hotel' to Launch in 2021.* Available at https://www.space.com/40207-space-hotel-launch-2021-aurora-station.html. Retrieved on 18 January 2019.

Space Daily. (2018a, June 27). *Japan Space Probe Reaches Asteroid in Search for Origin of Life.* Available at http://www.spacedaily.com/reports/Japan_space_probe_reaches_asteroid_in_search_for_origin_of_life_999.html. Retrieved on 18 December 2018.

Space Daily. (2018b, December 13). *Virgin Galactic Reaches Edge of Space in Historic Flight.* Available at http://www.spacedaily.com/reports/Virgin_Galactic_reaches_edge_of_space_in_historic_flight_999.html. Retrieved on 17 December 2018.

Swartwout, M. (2013). The First One Hundred Cubesats: A Statistical Look. *Journal of Small Satellites, 2*(2), 213–233.

The Conversation. (2018). *Mining Asteroids Could Unlock Untold Wealth—Here's How to Get Started* (May 2, 7:25 a.m. EDT). Available at http://theconversation.com/mining-asteroids-could-unlock-untold-wealth-heres-how-to-get-started-95675. Retrieved on 29 November 2018.

The Guardian. (2018, May 17). Space Tourism for the People: Become a Virtual Reality Astronaut. Available at https://www.theguardian.com/science/across-the-universe/2018/may/17/space-tourism-for-the-people-become-a-virtual-reality-astronaut. Retrieved on 18 January 2019.

The Verge. (2018, February 7). *Here's What's Next for SpaceX After Falcon Heavy's First Flight.* Available at https://www.theverge.com/2018/2/7/16983040/spacex-falcon-heavy-rocket-launch-schedule-spaceflight. Retrieved on 18 January 2019.

TMT. (2018). Official webpage online at https://www.tmt.org/page/about#what-is-tmt. Retrieved on 18 December 2018.

UN-World Space Week. (2018). Official Website. https://www.worldspaceweek.org/.

Wild, S. (2016). Giant SKA Telescope Rattles South African Community. *Nature, 534,* 444–446.

Wild, S. (2017). Giant Radio Telescope Scaled Back to Contain Costs. *Nature, 547,* 391–392.

9
Beyond Space Fiction

9.1 Exploring Our Neighbourhood

There are many scholarly contributions, which explore in great detail the advantages of space exploration, both in the Earth's neighbourhood and farther away, beyond the Solar System. Crawford highlights the benefits in the short terms as well as in a longer timeline. "The solar system is rich in energy and material resources which could potentially support a vibrant future space economy. Although many aspects of this economic activity will probably be pursued for purely commercial reasons (e.g. space tourism, and the mining of the Moon and asteroids for economically valuable materials), science will nevertheless be a major beneficiary. Indeed, science will benefit from all stages in the 'bootstrapping' of a space economy, from initial prospecting activities on the Moon and asteroids, through to the utilisation of the resulting resources to expand space activities. In particular, science will benefit from the infrastructure developed to support a space economy, which will help facilitate the construction of large space telescopes, the establishment of scientific research stations on the Moon and Mars (and perhaps elsewhere), and the mounting of ambitious space missions (including human missions) to the outer Solar System. In the

S. Paladini, *The New Frontiers of Space*,
https://doi.org/10.1007/978-3-030-19941-8_9

more distant future, an important scientific benefit of a well-developed space infrastructure may be the construction of interstellar space probes for the exploration of planets around nearby stars" (Crawford 2016: 10).

Among all the space activities and projects briefly itemised in Chapter 2, space exploration[1] is without any doubt the one that most resonates in the hearts and minds of people. When space advocates rationalise the efforts, both financial and technical, behind space missions, with economic motivations (as Crawford, quoted above), they only tell one part of the story and, by no means, the most important. The relevance of potential resource mining (as mentioned in the previous chapter) and other considerations are far more recent than the first missions sent to explore nearby planets. And if in those cases reasons of national prestige and the ongoing space race were powerful, they, again, did not justify the space exploration as a whole. Human curiosity and attraction for the unknown have always been present and in some cases represented the most important factor in space exploration.

It is impossible to give in one chapter a comprehensive picture of sixty years of activity in this area, and it is not the purpose here. What this section aims at doing is to illustrate the general principles of space exploration, showing the main areas of activities, the most important players, and the challenges and opportunities the world is facing at the present moment.

Section 2.2 in Chapter 2 has briefly covered milestones of the history of the space sector as a whole; therefore, here the data are sequentially presented by planet and celestial body. Table 9.1 gives instead a synthetic view of the year of the missions and the countries that have taken part in them.

The first observation is one that, compared to the countries that have managed to put a satellite in orbit, the number of space explorers is dramatically reduced. Since the 1950s, about 250 robotic spacecraft and 24 humans have left the Earth's atmosphere, and only a handful of countries are part of this elitist space club.

The reason is clear: compared to the satellite sector, exploration of outer space beyond the Earth's orbit requires at the same time a more substantial financial commitment, and effort sustained over many years, a certain technology prowess, and a vision of the role of the country in outer space. It is also important to notice that all the initial attempts

Table 9.1 Space explorations: Selected missions by countries

	USA (NASA)	USSR/Russia	ESA/EU	Japan (JAXA)	China	India
MOON	PIONEER ZERO (1958)	LUNA 2(1959), 9, 10 (1966)	SMART-1 (2003)	HITEN (1990)	Chang'e-3 (2007)	Chandrayaan-1/MOON IMPACT PROBE (2008)
	RANGER MISSIONS (1962–1964)	LUNOKHOD PROGRAMME (1969–1972)		KAGUYA (2007)	Chang'e-3/YUTU(2013)	
	APOLLO MISSIONS (1969–1972)					
	CLEMENTINE (1994)					
	LRO/LCROSS (2009)					
	EXPLORATION MISSION 2 (2023?)					
MARS	MARINER MISSIONS (1965–1971)	MARS 1M (1960), MARS 2 (1971)	EXOMAS/SCHIAPPARELLI (2016–ongoing)	NOZOMI (1998)	YINGHUO (2011)	MANGALAYAAN/MARS ORBITER (2014–ongoing)
	VIKING 1 (1976)	PHOBOS 1 & 2 (1988)				
	PATHFINDER/SEJOURNER (1997)					
	OPPORTUNITY (2003–ongoing)					
	CURIOSITY (2012–ongoing)					
	MAVEN (2013–ongoing)					
	INSIGHT (2018–ongoing)					
VENUS	MARINER 2 (1962)	HEAVY SATELLITE (1061)	VENUS EXPRESS (2006)	AKATSUKI (2010, 2015)	-	-
	PIONEER VENUS (1978)	VENERA 4 (1967), 5 & 6 (1969), 7 (1970), 9–10 (1975), 14 (1981)				
	MAGELLAN (1990–1994)	VEGA 1 & 2 (1985)				
MERCURY	MARINER 10 (1974)					
	MESSANGER (2011)					

(continued)

Table 9.1 (continued)

	USA (NASA)	USSR/Russia	ESA/EU	Japan (JAXA)	China	India
	BEPI-COLOMBO (2018–ongoing)					
ASTEROIDS	GALILEO (1991)			HAYABUSA 1 (2005–2010) 2 (2018)		
	NEAR SHOEMAKER (2000–2001)					
-VESTA	DAWN (2011)					
-CERES	DAWN (2015)					
-BENNU	OSIRIS-REX (2018)					
JUPITER	PIONEER 10 (1973)					
	PIONEER 11 (1974)					
	VOYAGER 1 & 2 (1979)					
	GALILEO (1995)					
	JUNO (2016-ongoing)					
SATURN	PIONEER 11 (1979)					
	VOYAGER 1 & 2					
	CASSINI (2004–2017)					
URANUS	VOYAGER 2 (1986)					
NEPTUNE	VOYAGER 2 (1989)					
PLUTO	NEW HORIZONS (2015)					
-KUIPER BELT	NEW HORIZONS					
COMETS	ISEE-3 (1985)					
	DEEP SPACE 1 (2001)					
	STARDUST (2006)					
-67P			ROSETTA/PHILAE (2014)			

Source Author's elaboration on official mission data from NASA; ESA; JAXA; other government sources (2019)

were military conceived and directed, while later missions were generally civilian-led (with possibly the exception of China, for the reasons explained in Chapter 5).

By looking at the timeline of the missions—more precise details have been given in appendices—it is also evident that space exploration depends on a series of conditions that have to be favourable, both on Earth at the moment of the launch (a few spacecraft, especially in the early days, have remained stranded or have exploded in orbit) and in terms of perigee and apogee[2] (relative distance of the target from the Earth).

This section will talk about the exploration of the Earth's neighbourhood, while the next one will cover the rest of the Solar System.

9.1.1 The Moon

Albeit being relatively near—375,000 km only—the Moon remains a place visited only by a handful of states (the USA, the USSR/Russia, China, India, Japan and by the European Space Agency). This speaks volume about the technical challenges space exploration presents.

The first-ever attempt to send a spacecraft into deep space aimed—logically enough—at the Moon. It was 1958, but this historic launch is not always mentioned in the official timelines of space exploration. The reason is that the US DoD-built PIONEER ZERO, launched six months after the first of all American satellite (Explorer 1) did not make it: its booster rocket exploded 73s after its launch (NASA 2019e, online).

It was another family of probes to land on another celestial object: LUNA.

LUNA 2 made it in 1959, after five unsuccessful attempts. Unsurprisingly, it was a Soviet achievement, like many others in that early period of the space race. LUNA 2 carried a few things over with it, such as Soviet emblems scattered on impact with the soil, sensors to detect a hypothetic magnetic field (not found), and even solid gas to enable tracking from the Earth. The Soviet perfected the attempt with the first automatic landing with LUNA 9 in 1966 and LUNA 10 (first artificial satellite on the lunar soil).

Fig. 9.1 The Apollo missions

It was only in the 1960s with the RANGER missions that the USA started the exploration of the Moon. RANGER 7 snapped the first image of the Moon by a US spacecraft on July 1964, before crashing down. RANGERs lay the groundwork in preparation of the Apollo programme, which managed to put a man on the lunar surface on 20 July 1969. So much has been written on the Apollo missions that it has not been covered here, apart from saying that, fifty years later, those missions still represent the only instances when a man has walked on another soil (Fig. 9.1).

The Moon remained a darling of space exploration, even though the USA, after EXPLORER 49 in 1973, abandoned the Earth's satellite for 21 years.

The Soviet LUNOKHOD programme in the 1970s deployed rovers on the lunar soil, which collected and returned samples back to Earth. The USA, with exception of the mission Clementine in 1994 only came back to the Moon in June 2009 with the launch of the LRO (Lunar Reconnaissance Orbiter), a spacecraft whose initial objective was to remain in orbit and map the lunar surface over a first phase of one year of duration. Since then, in its nine years of activities (and counting), LRO has delivered impressive science results, the most precise and complete topographic maps of the Earth's satellite to date, including a 3-D Map of the Moon at a 100-metre resolution and 98.2% coverage.

The third country to successfully reach the Moon was Japan, with its spacecraft HITEN in 1990 at its first attempt (Uesugi 2003). A successive series of missions followed in the 2000s, with KAGUYA, launching what was called the "undeclared Asian space race" to the Moon.

Chandrayaan-1 was the rather innovative way for India to reach the Moon in 2008, with an approach discussed in Chapter 5 and that has become an example for emerging economies. Together with the MOON IMPACT PROBE, the spacecraft carried payloads of other space agencies (such as NASA and ESA) and discovered the presence of water molecules in the lunar soil (*VOANews* 2009, online).

Apart from the USA and USSR, the only country to have "soft-landed" on the Moon is China, much in the same years of India and the Japanese KAGUYA.

After its first attempt to the Moon with Chang'e 1 in 2007, in December 2013 the spacecraft Chang'e 3 landed on the lunar soil and unleashed the lunar probe Yu-Tu (the Jade Rabbit) to collect samples and data. It was the first landing after LUNA 24 touched down in Mare Crisium in 1976 in its final mission of the programme. But it seems the Moon is on the space exploration's radar again, as mentioned in the conclusions.
China's last mission launched in December 2018: Chang'e 4 lunar lander is now exploring—to a detail never done before—the far side of the Moon.[3]

Apart from these five countries, the only other spacecraft to reach the Moon was launched by the European Space Agency. In 2003, ESA sent an orbiter (SMART-1) to collect imagery of the other missions and probes and test experimental technology, before crash landing 3 years later at the end of the mission.

But much more is to come, as the section about the space colonies and the so-called Moon Village shows in detail.

9.1.2 Mars

With the only exception of the Moon, Mars has been the most visited place by spaceships and, if actions will follow declarations, it will see the next human in spacewalking on its red, dusty surface. The hurdles, however, are formidable, and not just for crewed missions. There have been to date (December 2018) 44 robotic missions to Mars[4] from various space agencies, but only 19 of them have been successful. This translates into an astonishing failure rate of about 58%, making some people talking of a "Mars Curse" (CNN 2006).

Yet, Mars has been one of the first targets of space exploration efforts.

In 2015, NASA released a documentary (50 years of Mars exploration) to celebrate MARINER 4, the first US mission to Mars back in 1965, which managed the first fly-by. Many others have followed,[5] and some of them have made history, like MARS GLOBAL SURVEYOR in 1996 with the epic PATHFINDER, the first wheeled vehicle on the Martian soil. The descent of the probe was, very much as the Apollo 11, transmitted in world vision with millions of people following it, crashing the Internet browsing for images (NASA 2019e, online[6]). The new century has seen a flurry of missions to Mars, and NASA has been particularly active.[7] Among the flagship missions, the probe Curiosity in 2012, MAVEN in 2013 and, just in November 2018, INSIGHT.

As in the case of the Moon, the Soviet tried it first in 1960 with fly-bys and crash landing, before being beaten by the USA for the first successful mission (the already mentioned MARINER 4) and the first soft-landing on the Martian Sol (VIKING 1). The USSR succeeded partially in their missions (PHOBOS 1 and 2) to study Mars's moons Phobos and Deimos, returning imagery but failing to land probes.

Many other missions have ended prematurely, like the Japanese NOZOMI (1998), which failed to insert in the Martian orbits. After many essays, the mission was terminated in 2003, and the probe remained out there as a sort of artificial planet around the Sun just beside Mars's own orbit. China did not have better luck with its YINGHUO-1 (2011), stranded in the Earth's orbit. The Chinese have not got back to Mars so far, even though there are plans for a new mission in the 2020s.

Fig. 9.2 Mars rovers

ESA's ExoMars[8] managed to insert in orbit in 2016 but failed to land its SCHIAPPARELLI lander; only the first five minutes of data entry were retrieved.

This is why there was general amazement when India became in 2014 the first country ever to make it to the Red Planet at its maiden attempt, also becoming the first Asian nation to get there at all.[9]

There are at the moment two operational, moving rovers on the Mars (Opportunity and Curiosity), one stationary just landed (InSight, which, differently from other two, is drilling deep inside the surface to study the planet's composition) and six orbiters (MARS ODYSSEY, MARS EXPRESS, MARS RECONNAISSANCE ORBITER, MARS ORBITER, MAVEN and ExoMars).

The race to Mars is warming up (Fig. 9.2).

9.1.3 Venus

Among the celestial bodies of the immediate neighbourhood, Venus is the one who got less recent attention, at least after what its inhospitable surface has disclosed to human eyes. It has not been always like that: as a matter of fact, Venus was the first target of interplanetary fly-bys and landers, which contested well in the 1970s. In the case of Venus, most of the exploration has been carried out by the Soviet.

> The Soviet Venera program, from 1961 to 1985, is one of the largest efforts ever undertaken to study another planet. The fundamentals of interplanetary spacecraft design and remote sensing were first realized in the Venera missions. Successes included 3 atmospheric probes, 10 landings, 4 orbiters, 11 flybys or impacts, and 2 balloon probes in the clouds. Much of what is known today about Venus was discovered by these missions. (Mitchell 2018, online)

It was VENERA 1 (1961) the first probe sent to Venus—and to another planet whatsoever. However, the first fly-by was American (MARINER 2, 1962). VENERA 3 and 4 accomplished the first crash (1966) and soft landing (1967), while VENERA 7 returned the first photo of the Venusian surface (1970). The Soviet also managed the first Venusian orbit (VENERA 10, 1975), also continuing in the 1980s to explore the surface geology of the planet.

The USA sent two Pioneer missions to Venus (1978) to study the planet and they succeeded in 1994 (MAGELLAN, 1990–1994) to survey the entire surface of the planet.

Other two explorers have reached Venus: ESA with VENUS EXPRESS (2006), Japan with AKATSUKI (2010), which failed the insertion in orbit and remained in an alternative elliptic to image the surface. However, Venus is often used as a gravitational slingshot, and this means that several missions apart from the ones mentioned here have performed fly-bys, collecting data on their way to farther planets, which represents the topic of next section.

9.2 Farther Away: The Rest of the Heliosphere

The most evident thing when leaving Mars's orbit for looking at space missions farther away is the space agency behind them. For the overwhelming majority, all those efforts have been carried out by NASA, and only recently there have been initiatives from other players. But so far, it is not exaggerated to say that space exploration farther away the Earth's neighbourhood has been an American concept, and a NASA vision first.

9.2.1 Gas Giants and Their Moons

The logical reasons to treat the exploration of the four gas giants of the Solar System in the same subsection is due to the fact that at least two, if not all four, of them have been visited by the same spacecraft on its way to the edge of the system. Therefore, it is easier to talk about the missions themselves, since they are only a handful and all NASA's conceived and directed.

There are a set of constraints in reaching planets so far away, first of all, the energy cost that forces probes to approach planets for gravity assists through fly-bys and, more recently, to use ion thrusters as a way of propulsion. Another issue is the magnetosphere of the gas giants, together with the amount of radiation the shields of the probes have to sustain, which proved challenging for earlier missions (e.g. Pioneer).

Since the epic PIONEER 10, there have been only eight more probes that have reached the biggest gas giant to date. One of them, JUNO, is still out there, orbiting Jupiter and gathering data about its internal composition. This limited number plummets down to four when talking about Saturn, and to just one in the case of Uranus and Neptune (Fig. 9.3).

Moreover, among the eight missions to Jupiter, the majority where fly-bys[10] and only two missions to date managed an insertion into orbit (and no one ever landed; there's no solid surface to land). And if it was PIONEER 10 to set up a series of first in the exploration of the space beyond the Asteroid Belt ("the first vehicle placed on a trajectory to escape the solar system into interstellar space; the first spacecraft to fly beyond

Fig. 9.3 Juno

Mars; the first to fly through the asteroid belt; the first to fly past Jupiter; and the first to use all-nuclear electrical power", Siddiqi 2018: 45), it is to the Voyager missions that the actual understanding of the planet's moons and rings has come from.

If PIONEER 10 was the first probe to reach Jupiter, before heading to the edge of the Solar System, PIONEER 11 was the first to fly-by Saturn in 1979, even though the planet famous for its rings has been the target of one of the most famous-ever, and longest, exploratory missions: Cassini, a constant feature of the last decade in terms of science delivered and stunning imagery.

This is not the place to recapitulate all the discoveries made by the mission, which also performed the exploration of Saturn's moon Titan and showed its surface behind the dense atmosphere that surrounds its celestial body. But they literally changed the knowledge of the planet, providing surprising data about the moons themselves and discovering weird phenomena such as the hexagonal jet streams of Saturn's North Pole (about 20,000 miles wide and 60 miles into the atmosphere) and for which no definitive explanation has been found yet. In one of the fly-bys, Cassini even witnessed the birth of a new moon in 2014. Then, on Friday 15 September 2017, after 20 years in space, 13 of which spent

Fig. 9.4 Cassini

in Saturn's system, the probe plunged into the gas giant's atmosphere and was torn apart by Saturn's gravity in what has been called "the Gran Finale" (Fig. 9.4).[11]

9.2.2 Uranus and Neptune

"If you look up Uranus and Neptune in an encyclopaedia, there's a good chance the pictures you see will be about 30 years old" (The Planetary Society 2017, online). This is because there was only one single mission, and one probe, that reached the two farther gas giants of the Solar System, and both in the 1990s, albeit their orbits have been crossed by other spacecraft heading interstellar.

Of all the missions considered here, Voyager remains NASA's most ambitious. None ever visited, before and after, so many targets, nor provided insights as the twin spacecraft sent out in 1977. The two Voyagers took advantage of a planetary alignment happening only every 175 years to go for a "Gran Tour of the Solar System" and reaching the two so far unexplored planets. The missions' achievements are outstanding under any standard. "The two Voyagers have explored more planets (four), have discovered more moons (22), and have returned more photographic images, than any other spaceflight. The original price tag of nearly a billion dollars made it the second most expensive planetary voyage, exceeded only by Viking, which landed on Mars in 1976. Each Voyager spacecraft weighed more than any Surveyor or Ranger sent to the Moon and more than any Mariner or Pioneer probe (except for Pioneer Venus), though less than the combined weight of the Viking lander and orbiter" (NASA 1998, online).

Of the two Voyagers, Voyager 1 is even now NASA's farthest away and one of the fastest spacecraft ever built, launched days after its twin, on a faster trajectory to its mission to Jupiter, Saturn and Saturn's moon Titan. Then, it headed for the heliosphere, becoming the first probe ever to enter the interstellar medium in 2012.

Voyager 2's fly-by of Uranus occurred in January 1986 and Neptune in August 1989. Due to its speed and lacked the fuel, it didn't enter Neptune's orbit and headed interstellar (Murray 1989). NASA is now considering coming back, as the *Ice Giant Mission Planning* documents, released in 2017, show. "Exploration of at least one ice giant system is critical to advance our understanding of the Solar System, exoplanetary systems, and to advance our understanding of planetary formation and evolution" (NASA-LPI 2019i: 3). Taking into account propulsion methods and the availability of gravitational slingshots,[12] the launch windows are 2030–2034 time frame for Uranus missions and 2029–2030 for Neptune.

9.2.3 Comets, Asteroids, Dwarf Planets and Going Towards the Sun

There have been a few other missions worth of discussion in these last ten years.

The references mentioned at the beginning of the chapter will be of use to more in-depth studies, but here a few notes have been offered for analysis.

NEW HORIZONS (still ongoing) is one of them.

Started in 2006, and one of the fastest probes[13] to date, it was another iconic NASA's mission, aimed at exploring Pluto and the Kuiper Belt. The probe, after a trip of nine years, performed a historical Pluto's fly-by of 14 July 2015, returning stunning imagery and science, before heading to another dwarf planet (Ultima Thule). The object in the Kuiper Belt was reached by the probe in January 2019 (NASA-NEW HORIZONS 2019j).

ESA-Rosetta and Philae, and their historical achievements in 2014, have been covered in the case study in Chapter 8. Here, it is the case to say that it was not the first time a probe reached a comet, but the extent of the mission and the scientific results achieved by ESA were a first. The others, Hayabusa 1 (2010) and 2 (2018) and OSIRIS-REX (2018), have already been mentioned in the section relative to space mining.

Finally, the missions towards challenging celestial objects, one of which has just lifted off in summer 2018: BepiColombo. The ESA's latest probe represents Europe's first mission to Mercury, launched on 20 October 2018 towards the innermost and least explored terrestrial planet in the Solar System.

If everything goes as planned, it is due to arrive at destination in December 2025, after nine different planetary fly-bys, one of Earth, two of Venus and six of Mercury. The challenges are incredible; there's a reason it took until 1985 before somebody was able to figure out a way to make the orbital trajectories work out properly, and the spacecraft will have to manage in between a fast-orbiting planet and the immense gravitational pull of the Sun that threats to pull it out off course.[14]

9.3 Colonies

If there is an animated debate in the literature that deals with space exploration is the case for what should be the first step towards the colonisation of the Solar System. Because nobody doubts about the necessity to get there one day, albeit the enormous technical hurdles it will present for humanity. But, as Szocik explains, the necessity to solve the problem of decreasing resources that are needed for industry will eventually lead to this. "It is estimated that many of them will run out in the next fifty years.

Mars can be a place that could work as a source for new resources, perhaps more to process them on Mars than for sending them to Earth. However, the final solutions will depend on transport opportunities and their costs. An alternative solution for solving the problem of decreasing resources on Earth can be exploiting asteroids as the sources of metals that are inaccessible or just running out on Earth. We mean the concept of the mining of asteroids. Even if looking for new resources will not be the purpose of Martian colony, it might be an important part of this activity. One of the possible scenarios of this mission can be an opportunity to build the base from resources gained on Mars, for instance, by means of a 3D printer. However, even if we assume that looking for mineral and metal resources on Mars does not provide reasons for this mission, we can suspect that the scientific and technological advances needed for the mining of asteroids could be correlated with working on preparing manned mission to Mars, and it can work as one of the components of research and technological developments that are needed for this mission. The main benefit that could be provided by colonisation of Mars would be an opportunity to save the life of humanity when it is life on Earth will be endangered" (Szocik et al. 2017, online).

Space colony is a subject not only scientists and space operators have written about or the most. Science fiction writers—some of them scientists themselves[15]—have long debated and even envisioned it in a good deal of details. More recently, industry tycoon—Elon Musk is the most famous—have entered the fray and proposed their own version of Humanity 2.0. Interestingly, in Japanese popular culture, there are a few manga and anime that also tackled this subject, portraying which kind of consequences in terms of international relations this could end up with.[16]

At the present time, of all possible conceptual designs highlighted below, only the most basic—the Moon Village, as discussed Sect. 9.1.1—is technically feasible. Even this most basic version of a colony will demand a far more distributed and complex project than the ISS, which remains, however, the only useful example and guidance available to date. Moreover, for a certain degree of private initiative to take roots and become viable, it is necessary the development of a self-sustaining colonial economy, which produces in situ the goods it consumes and maybe even export them to the Earth. To make it happen, the commitment of governments' investments in the first phase of the colonies are indispensable, wherever they are going to be established.

To sum up different options, there are three main possibilities that have been discussed in concept studies:

- Colonies in a geostationary orbit.
- Moon colonies.
- Mars colonies.

Each one has its merits and not-so-hidden hurdles, which have been briefly outlined in what follows.

1. Colonies in Earth's orbit and/or asteroids[17]

A lot has been written about habitats revolving around the Earth, which can be all artificial or set over asteroids captured, put in orbit, and terraformed.

The idea of artificial orbiting habitat was first proposed in 1974 by Gerard K. O'Neill, professor of physics at Princeton University, who argued about their feasibility and economic sense, and wrote the first equations to prove them possible. He even speculated about the sociopolitical implications of a self-sustained colony (O' Neill 1974). Others proposed instead the capture of an asteroid to use as a habitat upon extensive terraforming.

The concept of terraforming[18] itself is fascinating, starting as a trope in science fiction (Williamson 2006), adopted by scientists in studies of space exploration and lately, even admitted into the English language (Brown 1993). It can be defined as "a process of planetary engineering, specifically directed at enhancing the capacity of an extraterrestrial planet, environment to support life. The ultimate in terraforming would be to create a planetary biosphere emulating all the functions of the biosphere of the Earth-one that would be fully habitable for human beings" (Fogg 1989: 23).

A few recent studies focused on the use of asteroids as military outposts in the context of planetary Cold War. Here, NOEs either are "turned into a kinetic energy weapon which can provide the same (or greater) threat as the nuclear arsenal of a present superpower" or they are "moved to Earth-centred orbits and utilised as 'solar power satellites (SPS)'".

2. The Moon

No matter if the Moon represented the first ever non-Earth soil humankind has walked onto, the satellite has long been visited only by rovers, even though, over the years, more states have joined the exploration, and it is actually based on this new trove of studies that the Moon has become again of interest for both states and private companies.

However, the concept for lunar colonies is not anything new: over the first half century of spaceflight has tended towards one of three general types: scientific base; mining settlement; lunar city. As it has been observed, "while useful as archetypes, single-purpose concepts are an impractical model for large-scale development of lunar activities. The most apt evidence for this conclusion is the historical contrast between vision and reality for Earth-orbiting space stations. Both Mir and Space Station Freedom were conceived as microgravity research laboratories whose development and utilisation was to be funded by single governments" (Sherwood 2017: 2).

And if the various US administrations have alternate interest to the Moon and to Mars as a target of the next stage of space exploration (Lucas and Hagan 2014), ESA has so far shown more interest in the Moon Village, intended to be a multinational outpost in space (Woerner 2015; Laguipo 2016). The Moon is considered here both as the basis for future space exploration and as a place for the establishment of a colony itself.

Adding to this renovated interest from governments and space agencies, there has been a series of initiatives in the private sector encouraging endeavours on the Moon, such as Google Lunar X-Prize supporting studies for operations (e.g. SpaceIL, Moon Express, Astrobotic) and the mining (e.g. Shackleton Energy Company, Space Resources) on the lunar soil. SpaceIL is scheduled to become, in 2019, the first privately funded probe to explore the Moon.

The literature has identified a series of factors which will affect the feasibility of such an endeavour, such as the potential activities it will have to perform; the scale of the project and its location on the lunar soil, which will have an impact on its environmental and operational constraint (Cohen and Benaroya 2009); potential participants and the relations among them; business models for growth and sustainability; and implications for the field of space architecture. Among the recent discoveries that have attracted interest in the Moon, one that is of particular relevance is the presence of ice under the lunar soil, both confirmed by the US's LMO and the Indian Chandrayaan-1. The availability of this essential component, which will at once provide fuel for rockets, drinkable water and components for fusion nuclear reactor (see Chapter 8 about the mining of raw materials in space) (Charania and De Pasquale 2007; Crawford 2015), has historically been one of the reasons for the choice of Mars over the Moon as the privileged location for a colony. Compared to both, ice on asteroids is technically more difficult—and therefore costly—to extract.

The real problem for an outpost on the Moon's surface is the comparable harsher environment compared to Mars, both in terms of gravity (1/6 on the Moon compared to Mars, which is about 30% of the Earth) and the lack of an even thin atmosphere (like the Martian one), which in turns makes much more complicated for the physiology of the human body to adapt to the more extreme temperatures and cosmic radiation. This is why, potential and vicinity notwithstanding, it is unlikely that the Moon Village would ever represent more than a scientific facility.

For a full-fledged colony, at least in the mid-term future, Mars looks more likely.

3. Mars

Mars is, not unsurprisingly, the place which features the most prominently as a possible target of space colonisation. It is easy to understand why; while there is a lot of other interesting alternatives for a space outpost, for the next 20–30 years all crewed missions in view of colonisation are aimed at Mars.

This is not a total novelty: Mars has appeared on the scientists' radar at the turn of the new century,[19] and, since that moment, it has never lost its appeal.

There are both scientific and technical reasons beyond that, some of which have already discussed above. Nonetheless, the hurdles will be formidable, and the comparative increase in distance from the Earth will force any Martian outposts to become a self-sustained system to a more comprehensive extent than the Moon.

The majority of Martian colonisation models previews some sort of terraforming; one of the main issues would be the warming up of the planet's surface, for which many methods have been proposed. The earliest ones were based on the now defunct "Long Winter Model" of the Martian climate (Sagan et al. 1973; Avemer and MacElroy 1971), which focused of releasing ice trapped in the Martian solar caps. Others (Birch 1992; Zubrin 1995) involved the use of orbiting mirrors to increase its amount of solar energy that reaches the surface.

The challenges are tremendous, at all levels. Shishko discusses the technicalities of a possible Mars colony's architecture, which is going to "contain a variety of infrastructure elements (systems), many of which will contain technologies that have yet to be developed. These systems would have to be successfully put into operation on Mars and sustained over long periods. Using current systems engineering terminology, one would describe a Mars Colony as a system-of-systems, but to truly grasp the magnitude of the undertaking, it is worthwhile to put any Mars Colony (and the systems that might comprise its architectural building blocks) in the context of its super-system. A Mars Colony exists within a complex that includes an interplanetary supply chain (and its component systems) and terrestrial enablers" (Shishko et al. 2017: 10).

That Mars makes people dream is a fact. It is not surprising that, more recently in time, a few private initiatives were launched towards Mars private exploration, like the Mars Society, the Dutch Mars ONE,[20] which started since 2013 the application process for perspective astronauts—and Elon Mask's SpaceX.

9.4 The Search for Alien Life: Hunting Exoplanets

Much of space exploration to date aimed at answering the fundamental question whether humankind is alone in the universe or not. The so-called Fermi Paradox that addresses the contradiction between the likelihood of the existence of alien life and the lack of evidence, has often been mentioned at this regard, together with the equation[21] that models it. However, the question should be more properly framed in terms of intelligent life than life *tout court*. This is because astrobiology principles dictate that when there are the basic conditions for life[22] there is also overwhelming possibilities of the existence of life itself, albeit primitive.

If this is the case, there are also fair odds the discovery of alien life next door is just decades ahead, with the outer moons of Jupiter and Saturn as the most promising candidates.

NASA put in a budget request in 2014 for a mission to be held in the 2020s, of about US$25 million. With the increased budget the agency received in 2016, the mission has been scheduled in. Europa Clipper is set to launch in June of 2023 and it will place a spacecraft in orbit around Jupiter to perform a detailed investigation of Europa's conditions. "During the nominal mission, the spacecraft will perform 45 flybys of Europa at closest-approach altitudes varying from 1700 miles to 16 miles (2700 kilometres to 25 kilometres) above the surface" (JPL-NASA 2018, online).

Europe is an icy world roughly the size of our Moon and whose features that hint at the possibility of alien life. First of all, it has geysers—spotted in 2013 by Hubble telescope, in the form of plumes of liquid on the surface—and, due to the tidal energy caused by Jupiter's immense gravity pressure, it hosts water in liquid form, in huge oceans under its frozen mantel.

The other two most likely candidates for alien life—Saturn's Enceladus and Titan—have not yet missions planned, but there are strong chances they are going to be explored too at some moment in the next two decades. "Confirmation that the chemical energy for life exists within the ocean of a small moon of Saturn is an important milestone in our search for habitable worlds beyond Earth. We now know that Enceladus has almost all of the ingredients you would need to support life as we know it on Earth" (NASA 2013, online).

The quest for something more complex than extremophile is still ongoing, though, and it has started long ago, with the same missions discussed in Sect. 9.3 and aimed at exploring the Solar System's gas giants. After carrying out their missions, a few of them have continued on their heliosphere crossing.

As of December 2018, there are currently four probes drifting in the interstellar space,[23] PIONEER 10, PIONEER 11, VOYAGER 1 and VOYAGER 2, two of them sent like "message in a bottle" to the discovery of alien life in the heliosphere and farther away.[24]

Of the four probes, Voyager 1 has entered, at some moment in 2012–2013,[25] the interstellar medium, while Voyager 2 crossed it only on 10 December 2018[26]; the other two are still in the heliosphere, on their way out.

The Deep Space Network (see Chapter 2) keeps track of all of them and monitors their signals. NASA lost contact with the Pioneers, first with PIONEER 11 in 1995, then in 2003 with PIONEER 10.[27] After having crossed the orbits of the other gas giants and reaching a distance of 67 AU from the Sun in 1997, PIONEER 10 was still able to transmit. The signal faded away—the last attempt was made in 2006—and by 5 November 2017, the inert PIONEER 10 spacecraft was roughly 118.824 AUs from Earth, a range second only to Voyager 1.

But there is still an ongoing Voyager Interstellar Mission (VIM) that will probably last until the 2020s, when the probes will run out of power to operate the scientific instruments on board and to send data back to Earth.

The latest test operated on Voyager 1[28] on 28 November 2017 was successful: the probe managed to fire up the thrusters it uses to orient itself so it can communicate with Earth, for the first time in 37 years (NASA 2017b, online).

Both in their 41st year of travel, as of 29 November 2018, Voyager 1 was 144.05 AU from the Sun, heading towards the constellation of Ophiuchus; Voyager 2 was instead at 119.32 AU on its way to the constellation of Pavo and now at the limit of the interstellar medium. Until they switch off,[29] they will keep collecting interplanetary and interstellar, fields, particles and waves (FPW) science data (Fig. 9.5).

During the same period that has seen the start of those missions, there has been also institutional support to contact initiatives with a hypothetical alien intelligence, the most well-known being SETI, a decade long project that investigate alien signals. Originally started with the investigation of narrow-band radio transmissions, now the SETI Institute uses the Allen

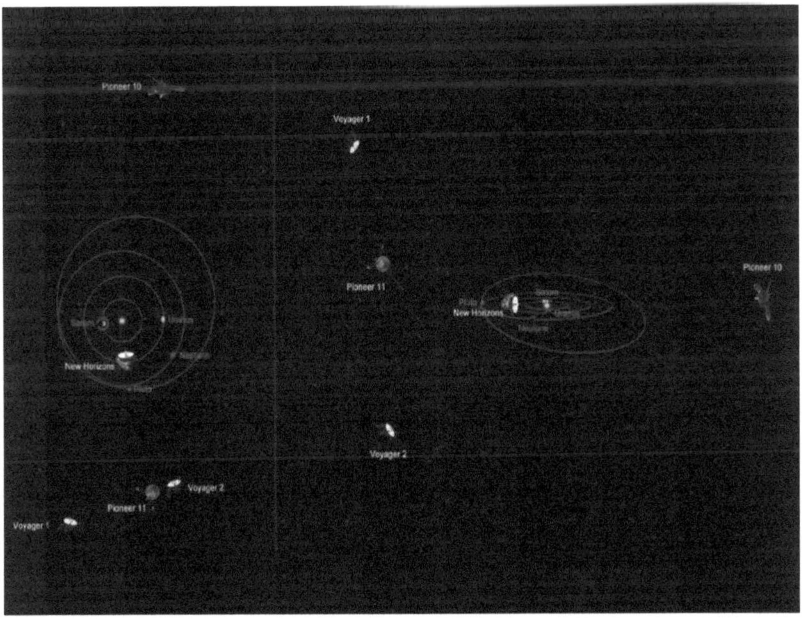

Fig. 9.5 (Estimated) position of the interstellar probes (as of December 2018) (This research has made use of the NASA Exoplanet Archive, which is operated by the California Institute of Technology, under contract with the National Aeronautics and Space Administration under the Exoplanet Exploration Programme.) (*Source* NASA)

Telescope Array (ATA) California, survey of tens of thousands of red dwarf stars.

The most recent in terms of initiatives for the search of alien life is the so-called exoplanet hunting: the efforts to discover and locate it in the nearest stars planetary systems and possibly rocky planets that can be good candidates for carbon-based life: Earth 2.0, as they have been soon dubbed.

The most famous planet-hunter is NASA's Kepler,[30] a space telescope (briefly mentioned in Sect. 8.1), whose photometre continuously monitored the brightness of a certain number of main-sequence stars. Kepler was launched in 2007 into an Earth-trailing heliocentric orbit, observing stars in the Milky Way and trying to make estimates about the number of exoplanets. During its eleven years of activity, it identified a stunning num-

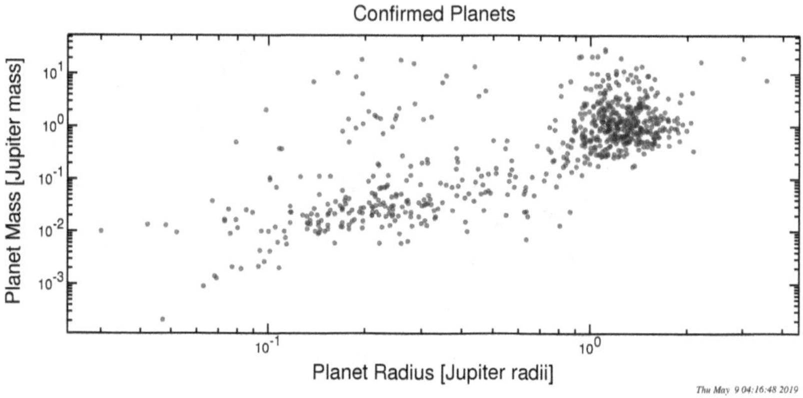

Fig. 9.6 Exoplanet statistics

ber of planets, in addition to even more planet candidates. Those numbers are still going up, as the incoming data are parsed, analysed, double-checked, and confirmed by other methods.[31] Since 1991, when the first exoplanet was identified, press news about discoveries became increasingly frequent, providing a long list of exoplanets in terms of shape, composition, and motion.

Exact figures vary: the catalogue by Exoplanet.eu[32] gives (cut-out date: 18 April 2019) 4040 planets in 3015 planetary systems, 658 of them in multiple planet systems.[33] NASA reports slightly different figures (with less confirmed planets—3940—but 3504 candidates to be validated) (Fig. 9.6).

Among this, many are rocky planets, i.e. planets with geological conditions similar to the Earth, even though not all of them are located in the Goldilock Zone.[34] Moreover, gas giants are of interests as well, because, like Jupiter and Saturn, they are likely to have a cohort of moons (so-called exomoons) that can be good candidates for life.[35]

Kepler ended its mission on 30 October 2018, but more is to come: Transiting Exoplanet Survey Satellite (TESS), TESS space telescope is already operational, launched on a two-year survey to map out exoplanets orbiting bright stars. The first batch of data, coming out of the observation of Sectors 1 and 2 between 25 July and 20 September 2018, has been already released and are available to the public through the online system MAST.[36]

From watchful artificial eyes on the sky to probe-assisted explorations, the search for exoplanets and alien life has just started.

Case Study: Imagining the Future—Science Fiction as a Roadmap for the Future of Humanity in Space

There are only a few fields where reality and fiction are as interconnected as in the space sector. This should not come as a surprise. "Since the middle of the twentieth century, over and above its contributions to science and technology, space activity has contributed to the production/ knowledge structure through down-stream technologies and public engagement with space-based products and services. Furthermore, looking through the lens of IPE power structures, we note that space activity has demonstrated its worth to the knowledge structure through arts, culture, film, television, literature and music" (Liberman 2017: 4).

Literature has been since the beginning one of the most influential, followed by movies and, more recently, by television, as this case illustrates.

Science fiction[37] has a long and powerful tradition, and, especially in the so-called golden years was a matter of interest not just for writers but for scientists, and it is not uncommon to have the very same people represented in both.

One of the most famous examples is Arthur C. Clarke, whose work[38] has won all the possible literary science fiction awards (e.g. Nebula, Hugo) and that was so esteemed by the science community to serve twice as the president of British Interplanetary Society (1946–1947; 1951–1953). Clarke was also the CBS commentator in occasion of the historical Apollo 11 Moon landing (and again for other two missions; McAleer 2012).

In some notable cases, devices first imagined by science fiction writers became later adopted by astrophysicists, as for the concept of the Dyson sphere. A Dyson sphere is a (still hypothetical, albeit some recent discussions[39]) structure that encompasses a star, capturing its energy. It has been named by the theoretical physicist Freeman Dyson, which, in turn, declared to have been inspired by Olaf Stapledon's 1937 science fiction novel *Star Maker*.

But science fiction also proved to be a mirror for the existing political tensions at the world's level, and no-one better than Robert A. Heinlein, a former US Navy officer, exemplifies it. Heinlein's novels portrayed many aspects of the Cold War, on Earth as much as in the space race; an example is his 1949 novel *The Long Watch*, which describes "an attempted military coup on the US Moon Base".

If a cultural equivalent of the 2010s has to be found, it could well be *Luna: New Moon* by the British author Ian McDonald. His corporations based on the Moon (the Five Houses that include families from Chinese, Brazilian and Russian ancestry made the readers think of BRICS in space) and the mercantilisation of every relationship, civil and even criminal, on the Earth's satellite, is a good mirror of the present times.

In the last decade, the popularity of science fiction has notably increased, for more than one reason. As noted by Dodds, "over that last ten years [from 1995 to 2005], disciplines such as International Relations (IR) have embraced the so-called 'return of culture' in a variety of ways. It has been raised not only as an issue of forthcoming insecurity in the form of civilisations and corresponding clashes, but also in the context of how popular culture might influence/shape/reproduce the foreign and security policies of governments" (Dodds 2006: 34). This has never been so evident when looking at space-related movies, fictions, literature and, more recently, with social medias.

King and Kryzwinska (2006) state that "science fiction dominates many of our movie screens and can be seen as a powerful cultural barometer of our times"; indeed, as they further note, it is a genre which "extends into a wider range of industrial, historical and cultural contexts".

Science fiction is an overarching term that includes more than it explains and for complex space operas, with or without aliens, set in other planetary systems or other galaxies altogether (Iain Banks, Peter F. Hamilton, Louis McBujoil, Steven Baxter, only to mention a few of them) there are others which address the Solar System instead and medium-term future.

A few authors, especially in the early days, explored near-Earth habitats (some of which discussed in this chapter), as Ben Bova in his 1978 novel *Colony.* The habitat, Island One, represents here the only refuge from the world, orbiting at the Earth-Moon L4[40] point. Another example is John C. McLoughlin's space-habitat, an asteroid where humanity has escaped to after a doomsday on Earth, described in 1983 in *The Helix and the Sword.*

More recently, others have been looking at the life on the other planets from the point of view of the many challenges colonies in the Solar System would present for the Earth, in a sort of re-enactment of Wallerstein's theory of "the core and the periphery" that emerged in the Modern World in the 1500s.

The main difference with science fiction of the early days is that the focus has shifted from technological to cultural aspects and political implications (e.g. Corey's The Expanse, whose first book was Leviathan Wakes, 2011, now a Netflix series), even though some authors do a good job in examining both. A good example is Kim Stanley Robinson, whose famous trilogy about colonising (*Red Mars* 1992), terraforming (*Green Mars* 1993), and finally inhabiting over the long term (*Blue Mars* 1996) the Red Planet is a classic.

Finally, non-Western views.

A lot has been done by the science fiction community in the last decade to open to non-Western views of space. If anything, the success of a novel such as "the Tree Body Problem" by the Chinese Liu Cixin is a good indication.

But there is another example, even more iconic if possible, because it has been around for forty years (2019 will be the fortieth anniversary) and still represents one of the most famous non-Western world of the humanity in space: the Japanese franchise Gundam (ガンダム).

Created by the Japanese mangaka Yoshiyuki Tomino (富野 由悠季) and produced by Sunrise Animation Studio, the original anime series was first aired in 1979. From there, it went on generating many sequels, prequels, spin-off (44 up to 2019, only considering the anime), manga, OVAs, video games and an impressive range of merchandise, for about US$25 billion in revenues[41] (Comicbook.com 2018).

For longevity, success, and iterations, Gundam has been compared to the US-created Star Trek, but, differently from Star Trek, it does not feature aliens and space exploration of faraway galaxies. The centre of action is the Earth and its colonies—artificial habitats situated on the Earth–Sun Lagrangian points, which at a certain moment decide to rebel and obtain their independence from the Earth Command. While the series feature giant robots (mecha) and powerful spaceships, the emphasis is on the contraposition between Earth-born people and spacenoids, whose incapability to comprehend any longer fellow humans "whose soul is weighed down by gravity" eventually leads to a never-ending war.

A cautionary tale for the future of mankind in space.

List of References

Dodds, K. (2006). Screening Geopolitics: James Bond and the Early Cold War Films (1962–1967). *Geopolitics, 10*(2), 266–289.

The Guardian. (2018, January 3). No Alien Megastructure Around Mysterious 'Tabby's Star', Analysis Shows. Retrieved 7 January 2019, from https://www.theguardian.com/science/across-the-universe/2018/jan/03/no-alien-megastructure-around-mysterious-tabbys-star-analysis-shows.

King, G., & Kryzwinska, T. (2006). *Science Fiction Cinema: From Outer Space to Cyberspace*. London: Wallflower.

McAleer, Neil. (2012). Sir Arthur C. Clarke: Odyssey of a Visionary—A Biography. Clarke Project.

'Pokemon' Is the Highest-Grossing Franchise of All-Time. (2018). Anime, 23 June 2018. Retrieved 19 January 2019, from https://comicbook.com/anime/2018/06/24/pokemon-franchise-revenue-anime-media/.

Lieberman, S. (2017). Understanding Our Changing Relationship with Space: An International Political Economy Reading of Space Popularisation. *Space Policy, 41*, 53–59.

Notes

1. The essential reference for this and the following section is Siddiqi (2018), which covers sixty years of space exploration. While being a NASA-published book focuses especially on American mission, it is still the master database for any kind of serious analysis on this subject, together with the reports from the individual space agencies.
2. A quick look at the data from Mars launching will show the window of opportunity that happens every two years, and which is when the probes are launched.
3. This is the website of CNSA, China's national space agency: http://www.cnsa.gov.cn/n6443408/index.html.
4. Over 49 launches, 24 of them between up to 1970s (*Source* JPL/NSA).
5. NASA has an impressive footage of images and videos about Mars exploration and the details of all missions flying there. Soon after MARINER 4, MARINER 6 and 7 have followed in 1969, and MARINER 9 managed a successful insertion in orbit in 1971. VIKING 1 and 2 soft-landed in 1976, remaining, respectively, operational for six and three years.
6. JPL website got, back in those Internet's early days, 30 million hits on the landing day (4 July 1997) and about 80 in the weekday after.
7. Among the successful ones, there are to be remembered MARS ODYSSEY in 2001, other rovers in 2003, MARS RECONNAISSANCE ORBIT IN 2005 and MARS PHOENIX LANDER IN 2008.
8. The spacecraft was a cooperation with Russian Roscosmos.
9. And with the least expensive budget ever. See Chapter 5 for more about the peculiar Indian space development concept.
10. It has been observed (Siddiqi 2018) that 'flybys" is not the correct words for this kind of missions, since spacecraft do not fly. It would be more correct to talk of swing by. However, since this is a remnant of the era of aeronautics still in current use by both press and space agency, they have been kept here.
11. NASA has made available all data and imagery about Cassini on the official webpage https://www.nasa.gov/mission_pages/cassini/spacecraft/index.html.
12. "The travel time to Uranus, using an Atlas V with five solid rocket boosters, is 12 years for an orbiter. It takes 13 years to get to Neptune using a Delta IV Heavy launcher--and that's with an extra solar-electric propulsion stage to add additional thrust. All missions require Jupiter gravity

assists, as well as likely flybys of Venus and Earth" (The Planetary Society 2017, online).

13. There are many claims about the fastest spacecraft—Voyager 1, New Horizons, Juno—but it all depends on the parametres and towards what it is measured [Astronomy (2016) Was Juno the fastest spacecraft ever? Only kind of. 12 July 2016, http://www.astronomy.com/news/2016/07/juno-may-not-have-been-the-fastest-spacecraft-ever]. Solar Probe Plus is scheduled to launch in July 2018, daring to skim closer to the Sun than any spacecraft before. When it does, it'll set all new records with a velocity of over 200,000 meters per second (450,000 mph) relative to the Sun. And then, it truly will be the fastest spacecraft, no matter how you measure it.
14. Otherwise said, instead of gain speed, the probe has to slow down (to know more about the technical feature and the engine that is going to deliver it, see BepiColombo's technical description here: https://newatlas.com/bepicolombo-mercury-ion-thruster/56878/.
15. A case study has been devoted to this important topic at the end of this chapter. Among the writers/scientists that more have contributed to this area, Arthur C. Clarke and David Brin. For a daring vision of a full-fledged colonisation of the Solar System, the writer Kim Stanley Robinson is a first reference.
16. The most obvious reference here is the Gundam franchise, which can be considered the non-Western alternative to Star Trek and that has also been studied at an academic level.
17. Here, asteroids include both NEOs and the celestial bodies in the Asteroid Belt, now a popular feature of the science fiction series "The Expanse" (see case study the end of this chapter).
18. Terraforming as a process is of course not restricted to asteroids, and as a matter of fact, it has been proposed for Mars back in the 1970 (Bums and Harwit 1973; Sagan 1973).
19. "At the present time, all research into planetary engineering, whether applied to Mars or anywhere else, is concerned entirely with defining the boundaries of the possible, rather than in charting some definite route into the future. The concept can no longer be described as fantasy, although confirmation of its practicality awaits a detailed exploration of Mars, an inventory of its resources, a better understanding of the phenomenon of planetary habitability, and a future where the solar system is opened to technological civilization as a new and expanding frontier" (Fogg 1998: 7).

20. Mars ONE will probably not go anywhere, since it has filed for bankruptcy in 2019. Before that, the non-profit organisation was however very active in terms of recruiting and media presence, with a blog and captivating website. See https://www.mars-one.com/about-mars-one/current-mission-status.
21. It is the Drake equation, first proposed by the astrophysics Frank Drake in 1961, which takes the simple forms of takes this simple form: $N = R^* \times f_p \times n_e \times f_l \times f_i \times f_c \times L$.

    For a discussion on this fascinating point, see https://www.space.com/25219-drake-equation.html.
22. There are a lot of discussions here, but for what the kind of life humans are capable to recognise as such—i.e. carbon-based—a few conditions have been identified: essential components (the primordial soup) liquid, enough time in geological terms to develop.
23. Five in including NEW HORIZONS, which, however, is still on its mission. This is why has not been included here. It is also the case to say that PIONEER 10 is heading in a completely different direction from the other historical missions. "The spacecraft is generally heading in the direction of the red star Aldebaran which forms the eye of the Taurus constellation. It is expected to pass by Aldebaran in about two million years. Pioneer 10 is heading out of the solar system in a direction very different from the two Voyager probes and Pioneer 11, i.e., towards the nose of the heliosphere in an upstream direction relative to the inflowing interstellar gas. In case of an intercept by intelligent life, Pioneer 10 carries an aluminum plaque with diagrams of a man and a woman, the solar system, and its location relative to 14 pulsars. The expectation is that such intelligent life would be able to interpret the diagram to determine the position of the Sun (and thus, Earth) at the time of launch relative to the Pulsars" (Siddiqi 2018: 67).
24. If this is indeed a good idea or not, it is a completely different question. Stephen Hawking, among others, has often cautioned against getting into touch with alien intelligence.
25. Scientists believe it entered interstellar space or the space between stars, on 25 August 2012. However, there is not a definite line, but more of a zone where "termination shock" from the Sun stops, the so-called heliopause Space.Com (2013).
26. NASA announced that Voyager 2 entered interstellar space on 10 December 2018 (NASA Press Release 2018).

27. In 2016, NASA estimated PIONEER 10 to be approximately 10 billion miles from Earth, travelling 26900 mph. In case it were not hit or destroyed by space debris, it has been just surpassed on its way to the interstellar by the still active Voyager 2.
28. All of Voyager's thrusters were developed by Aerojet Rocketdyne. The same kind of thruster, called the MR-103, flew on other NASA spacecraft as well, such as Cassini and Dawn. (NASA 2017a). A test for Voyager 2 is to be performed soon, even though there are less concerns in this case than for the other probe and the results should be satisfactory.
29. New Horizons is another mission that is going to provide precious science of the medium, as well. "Due purely to serendipity, Pluto is now located toward the nose of the heliosphere. Although this mission will not return data from the interstellar medium, it will measure the pickup ions directly, providing new information on the densities and composition of the interstellar neutrals entering the solar system" (Schwadron 2017: 459).
30. NASA maintains a database with Kepler's discoveries at: http://kepler.nasa.gov/.
31. There are a few ways to detect exoplanets, for example, microlensing, transits, or infrared, and they're often used together. For example, in the case of two of 2018 discovery, NASA "added two new planets that were discovered using the microlensing technique. The first planet, UKIRT-2017-BLG-001L b is the first microlensing planet to be detected solely in the infrared. The second planet, OGLE-2017-BLG-0173L b, is a super-Earth that is similar to the K2-18 system, where the b planets have similar Earth masses and orbit very closely to their parent stars, making comparisons of these two systems potentially very interesting" NASA (Press Release, 26 March 2018).
32. Exoplanet.EU is an open web repository, maintained by Exoplanet Team at the Observatoire de Paris—LESIA, Exoplanet.EU.
33. The tables can be downloaded at http://exoplanet.eu/catalog/ and filtered by categories.
34. Also called Habitable Zone, as it is the range where conditions are considered the best for developing life.
35. While their size makes the localisation of these exomoons incredibly difficult, one has been recently discovered in 2018.
36. The Web archive and the access are obtainable here: https://mast.stsci.edu/portal/Mashup/Clients/Mast/Portal.html.Among its main features, thousands of astronomical data archives from around the world for

images, spectra and the Hubble Source Catalogue with a hundred million measurements of objects in Hubble images.

37. The present section has no ambition to present the state of science fiction today and even less its history. Even an entire monograph will not be enough for it. The idea here is to provide some hints on the way science fiction has imagined since the start of the space age humankind in outer space, putting it in relationship with cultural and political aspects. Some references for further reading have been provided at the end and, again, have no ambition to be exhaustive. They represent merely a starting point according to the author of this book's very personal (and therefore biased) views.
38. Some of Clarke's novels remain among the most read and influential ever in the history of science fiction.
39. In 2015, there was a lot of fuss in the field for the discovery of an unusual dimming of a star KIC 8462852, also called "Tabby Star". Nobody could explain its behaviour and there were wide speculations about it—one being the existence of a "Dyson sphere" around it. Successive analyses proved it untrue (*The Guardian*, 3 January 2018.
40. L4 stays for Lagrangian points, and they are five special orbital positions where a body between two larger masses orbits in a constant pattern. NASA and other space agencies regularly use the ones on Sun–Earth orbit. For example, L1 point allows an uninterrupted view of the sun, and therefore, the Solar and Heliospheric Observatory Satellite (SOHO) is set there. These points are quite famous in science fiction and they have been considered as privileged locations for Earth's colonies. In O'Neill's words "for the first colony it is probably best to choose a particular point on that sphere, within easy range of both Earth and Moon, not so close as to be eclipsed often, and preferably stable against displacements in all three coordinates. The L4 and L5 satisfy all these conditions" (O'Neill 1974, online).
41. In the list for the world's highest-grossing media franchises, Gundam came out 15th, after Star Wars (US$65 billion) but before Star Trek, which "only" grossed an estimated US$10.1.

References

Avemer, M. M., & MacEiroy, R. D. (1971). *On the Habitability of Mars: An Approach to Planetary Ecosynthesis*. NASA SP-414. Washington, DC: NASA.

Birch, P. (1992). Terraforming Mars Quickly. *Journal of the British Interplanetary Society, 45*, 331.

Brown, L. (1993). The New Shorter Oxford English Dictionary, Vol. 2 (N–Z). Oxford: Clarendon Press.

Bums, J. A., & Harwit, M. (1973). Towards a More Habitable Mars-or-the Coming Martian Spring. *Icarus, 19*, 126–140.

Charania, A. C., & De Pasquale, D. (2007, September). *Economic Analysis of a Lunar in Situ Resource Utilization (ISRU) Propellant Services Market*. 58th International Astronautical Conference—IAC Paper 07–A5.1.03.

CNN. (2006). *Beating the Curse of Mars*, 23 December. http://www.cnn.com/2006/TECH/science/12/23/mwonders.mars/index.html. Retrieved on 7 January 2019.

Cohen, M. M., & Benaroya, H. (2009). Lunar-Base Structures. In A. S. Howe & B. Sherwood (Eds.), *Out of This World: The New Field of Space Architecture* (pp. 179–204). Reston, VA: AIAA, Chapter 15.

Correll, R. R., & Peter, N. (2005). Odyssey: Principles for Enduring Space Exploration. *Space Policy, 21*(4), 251–258.

Crawford, I. A. (2015). Lunar Resources: A Review. *Progress in Physical Geography, 39*(2), 137–167.

Crawford, I. A. (2016). The Long-Term Scientific Benefits of a Space Economy. *Space Policy, 37*, 58–61.

ESA. (2019). *Organisation Website*. Our Activities. Available at http://www.esa.int/Our_Activities/Space_Science/SMART-1. Retrieved on 7 January 2019.

Fogg, M. J. (1989). The Creation of an Artificial Dense Martian Atmosphere: A Major Obstacle to the Terraforming of Mars. *Journal of the British Interplanetary Society, 42*, 57.

Fogg, M. J. (1998). Terraforming Mars: A Review of Current Research. *Advances in Space Research, 22*(3), 415–420.

Forganni, A. (2017). The Potential of Space Tourism for Space Popularisation: An Opportunity for the EU Space Policy? *Space Policy, 41*, 48–52.

Futurism. (2018). China and Europe May Build a "Moon Village" in the 2020s. *The Moon Could Be Earth 2.0*. Available at https://futurism.com/china-and-europe-may-build-a-moon-village-in-the-2020s. Retrieved on 7 January 2019.

JPL-NASA Deep Space. (2018). Deep Space Network—NASA Jet Propulsion Laboratory Homepage. Available at https://deepspace.jpl.nasa.gov/. Retrieved on 27 March 2018.

Kecskes, C. (2002). Scenarios Which May Lead to the Rise of an Asteroid-Based Technical Civilisation. *Acta Astronautica, 50*(9), 569–577.

The Guardian. (2018). *No Alien Megastructure Around Mysterious 'Tabby's Star', Analysis Shows.* Available at https://www.theguardian.com/science/across-the-universe/2018/jan/03/no-alien-megastructure-around-mysterious-tabbys-star-analysis-shows. Retrieved on 1 January 2019.

Laguipo, A. (2016). ESA to Build 3D-Printed Moon Village for Astronauts by 2030. *Tech Times.* Available at http://www.techtimes.com/articles/121906/20160107/esato-build-3d-printed-moon-village-for-astronauts-by-2030.htm. Retrieved on 10 January 2017.

Lucas, M. T., & Hagan, P. C. (2014). Comparison of Two Excavation Systems for the Mining of Lunar Regolith. *Journal of Research Projects Review, 3*(1), 39–44.

Mars Daily. (2014). India's Mars Orbiter Cost Only 11 Percent of NASA's Maven Probe: Reports, 26 September 2014. Available at http://www.marsdaily.com/reports/Indias_Mars_Orbiter_Cost_Only_11_Percent_of_NASAs_Maven_Probe_Reports_999.html. Retrieved on 18 December 2018.

Mentallandscape.com. (2018). Venera: The Soviet Exploration of Venus. Available at http://mentallandscape.com/V_Venus.htm. Retrieved on 7 December 2019.

Michtell, Dan. (2018). *Venera: The Soviet Exploration of Venus.* Available at http://mentallandscape.com/V_Venus.htm. Retrieved on 28 November 2018.

Murray, B. (1989). *Journey into Space: The First Three Decades of Space Exploration.* New York, NY: W. W. Norton.

NASA. (1998). Voyager: The Grand Tour of Big Science. In *From Engineering Science to Big Science.* https://history.nasa.gov/SP-4219/Chapter11.html. Retrieved on 7 January 2019.

NASA. (2013). *Cassini Spacecraft Reveals Forces Controlling Saturn Moon,* 31 July 2013. Available at https://www.nasa.gov/mission_pages/cassini/main/index.html. Retrieved on 17 January 2018.

NASA. (2017a). *Mars Rover Tests Driving, Drilling and Detecting Life in Chile's High Desert.* Available at https://www.nasa.gov/feature/ames/mars-rover-tests-driving-drilling-and-detecting-life-in-chile-s-high-desert. Retrieved on 10 February 2018.

NASA. (2017b). *Voyager 1 Fires Up Thrusters After 37 Years December 1, 2017*. Available at https://www.nasa.gov/feature/jpl/voyager-1-fires-up-thrusters-after-37. Retrieved on 17 January 2018.
NASA. (2019a). *Explore Solar System And Beyond*. Available at https://www.nasa.gov/topics/solarsystem/index.html. Retrieved on 7 January 2019.
NASA. (2019b). *LUNA 2*. Available at http://sse.jpl.nasa.gov/missions/profile.cfm?Sort=Target&Target=Moon&MCode=Luna_02.
NASA. (2019c). *LUNA 10*. Available at https://solarsystem.nasa.gov/missions/luna-10/in-depth/. Retrieved on 7 January 2019.
NASA. (2019d). *Outer Space Mission*. Available at https://solarsystem.nasa.gov/missions/?order=launch_date+desc&per_page=50&page=0&search=&fs=&fc=&ft=&dp=&category=. Retrieved on 7 January 2019.
NASA. (2019e). *PIONEER ZERO*. Available at https://solarsystem.nasa.gov/missions/pioneer-0/in-depth/. Retrieved on 7 January 2019.
NASA. (2019f). *LRO (Lunar Reconnaissance Orbiter)*. Available at https://lunar.gsfc.nasa.gov/. Retrieved on 7 January 2019.
NASA. (2019g). *NASA Memorandum: 1958 NASA/USAF Space Probes (Able-1) Final Report*. Available at https://www.sdfo.org/stl/1958%20NASA%20USAF%20Space%20Probes%20%28ABLE-1%29%20FINAL%20REPORT%20Volume%2002.pdf. Retrieved on 7 January 2019.
NASA. (2019h). *NASA-VIM*. Available at https://voyager.jpl.nasa.gov/mission/science/. Retrieved on 7 January 2019.
NASA. (2019i). *LPI—Ice Giant Mission Study Final Report*. Available at https://www.lpi.usra.edu/icegiants/mission_study/Exec-Summary.pdf. Retrieved on 7 January 2019.
NASA. (2019j). *NEW HORIZONS*. Available at http://pluto.jhuapl.edu/. Retrieved on 7 January 2019.
NASA. (2019k). *50 Years of Mars Exploration*. Available at https://www.nasa.gov/mission_pages/phoenix/main/index.html. Retrieved on 7 January 2019.
NASA. (2019-l). *The Day the Internet Stood Still*. Available at https://www.nasa.gov/specials/pathfinder20/. Retrieved on 7 January 2019.
The New York Times. (2018). China's Chang'e-4 Launches on Mission to the Moon's Far Side, 7 December 2018. Available at https://www.nytimes.com/2018/12/07/science/china-moon-change-4.html. Retrieved on 18 December 2019.
Neal, C. R. (2009). The Moon 35 Years After Apollo: What's Left to Learn? *Chemie der Erde, 69*, 1–41.

O'Neill, G. K. (1974). The Colonization of Space, in *Physics Today*, in National Space Society. (2019). Available at https://space.nss.org/the-colonization-of-space-gerard-k-o-neill-physics-today-1974/. Retrieved 10 January 2019.

Park, M. (2014). India's Spacecraft Reaches Mars Orbit...and History. *CNN*, 23 September 2014. Available at http://edition.cnn.com/2014/09/23/world/asia/mars-india-orbiter/index.html. Retrieved on 10 February 2018.

Sagan, C. (1973). Planetary Engineering on Mars. *Icarus, 20,* 513–530.

Sagan, C., Toon, O. B., & Gierasch, P. J. (1973). Climatic Change on Mars. *Science, 18*(1), 104–125.

Schwadron, E. (2017). *The Limits of Our Solar System.* Available at ftp://space.mit.edu/pub/plasma/publications/jdr_schwadron_book/jdr_schwadron_book.pdf. Retrieved on 1 January 2018.

Schwamb, M., et al. (2012). Planet Hunters: Assessing the Kepler Inventory of Short-Period Planets. *Astroph Journal, 754*(2), 129–145.

Sherwood, B. (2017). Space Architecture for Moon Village. *Acta Astronautica.* https://doi.org/10.1016/j.actaastro.2017.07.019. Retrieved on 7 January 2019.

Shishko, R., Fradet, R., Do, S., Saydam, S., Tapia-Cortez, C., Dempster, A. G., et al. (2017). Mars Colony in Situ Resource Utilization: An Integrated Architecture and Economics Model. *Acta Astronautica, 138,* 53–67.

Siddiqi, A. A. (2018). *Beyond Earth: A Chronicle of Deep Space Exploration, 1958–2016.* NASA History Program Office. Available at https://www.nasa.gov/connect/ebooks/beyond_earth_detail.html. Retrieved on 7 January 2019.

The Atlantic. (2013). China Becomes Third Country to Ever 'Soft-Land' on the Moon. Available at https://www.theatlantic.com/international/archive/2013/12/china-becomes-third-country-ever-soft-land-moon/356151/.

The Planetary Society. (2017). *Revisiting the Ice Giants: NASA Study Considers Uranus and Neptune Missions,* 21 June 2017. Available at http://www.planetary.org/blogs/jason-davis/2017/20170621-revisiting-ice-giants.html. Retrieved on 7 January 2019.

Space.Com. (2012). *NASA Probe Beams Home Best Moon Map Ever.* Available at https://www.space.com/13666-moon-map-lunar-reconnaissance-orbiter.html.

Space.Com. (2013). *Voyager 1 Really Is in Interstellar Space: How NASA Knows,* 13 September 2013. Available at https://www.space.com/22797-voyager-1-interstellar-space-nasa-proof.html. Retrieved on 17 January 2018.

Space Daily. (2018). Japan Space Probe Reaches Asteroid in Search for Origin of Life, 27 June 2018. Available at http://www.spacedaily.com/reports/

Japan_space_probe_reaches_asteroid_in_search_for_origin_of_life_999.html. Retrieved on 18 December 2018.
Szocik, K., Wójtowiczb, T., & Baran, L. (2017). War or Peace? The Possible Scenarios of Colonising Mars. *Space Policy, 42*, 31–36.
Uesugi, K. T. (2003). Space Engineering Spacecraft (MUSES) Program in ISAS Featuring Its Latest Mission 'HAYABUSA'. *International Conference on Recent Advances in Space Technologies, 2003.*
VOANews. (2009). *Lunar Missions Detect Water on Moon*, 2 November. Available at https://www.voanews.com/a/a-13-2009-09-24-voa46-68806852/362715.html. Retrieved on 7 January 2019.
Williamson, M. (2006). *Space: The Fragile Frontier.* Reston, VA: AIAA.
Woerner, J. (2015, October 13). *ESA Director General Jan Woerner Meets the Press.* International Astronautical Federation Global Networking Forum, 66th International Astronautical Congress, Jerusalem.
Zubrin, R. (1995). The Economic Viability of Mars Colonization. *Journal of the British Interplanetary Society, 48*, 407–430.

10
Conclusions

2018 has been a record-breaking year for space activities, with more than 100 successful launches, the testing of a new powerful rocket (SpaceX's Falcon Heavy), new space exploration missions launched (e.g. BepiColombo, Mars InSight) and others making the news with some daring results (e.g. Osiris-Rex, Hayabusa 2, Chang'e 4). The ISS life has been extended to mid-2020s, a fact that is going to allow the development of privately funded stations like the already-deployed, inflatable Bigelow module.

2019—the 50th year anniversary of the first man on another celestial body—promises to be even better with what is already on the plans in terms of missions, launches, and the ongoing space race. As a matter of fact, it has already started delivering on those promises, at all levels. It began with a resounding success, New Horizons reaching Ultima Thule in the Kuiper Belt and performing a fly-by of the dwarf planet on 1 January. Also in January, China has got back to Moon with a historical landing on the Far Side—in the first mission ever on that side of the Moon—where it put a rover and even tried to grow crops (cotton) out there. In April, for the first time, a team of scientists was able for the first time ever to take a picture of a black hole, stunning the entire world.

More it is to come.

S. Paladini, *The New Frontiers of Space*,
https://doi.org/10.1007/978-3-030-19941-8_10

2019 should be the year where the USA comes back to ship astronauts to the ISS. NASA awarded its Commercial Crew Programme to two companies. One is SpaceX, which looks at launching its first crewed flight (Crew Dragon Demo 2, DM-2) in June from Cape Canaveral, Florida. This will be the first launch of astronauts from the US soil since the final Space Shuttle flight on 8 July 2011. The other is Boeing, which, upon successful testing of its CST-100 Starliner, will try a crewed flight in the second half of the year. Blue Origin is also looking at launching a crewed flight in 2019, but there is no date fixed yet. Virgin Galactic aims instead at taking its first paying customers to suborbital flights after the success of December 2018.

The growth of commercial launches has already proved of keeping pace in 2019, together with the private operators in the field. The NZ-based Rocket Lab aims at scaling up its operations after its first two launches in 2018. Its rocket, Electron, is manufactured to cater to the needs of CubeSat and small satellites, and according to the company's plans, it should be able to make up to 120 launches in the year.

Another company, Vector Launch has the same target, and it will test its smallsat rocket Vector-R later in the year. ISRO, the Indian Space Agency, is also developing the Small Satellite Launch Vehicle (SSLV), which should be ready by the end of the year.

For what concerns space exploration, a few other missions are on the way.

India is looking to get to the Moon for the second time with Chandrayaan-2, which will consist in an orbiter, a lander and a rover. The plan is to explore the Moon's South Pole.

As a reminder of how challenging space exploration is, in April 2019 the first privately sent Moon lander crashed on the lunar soil. The non-profit Israeli firm, SpaceIL, had managed to send its Moon-bound spacecraft, Beresheet, on a SpaceX Falcon 9, to explore the lunar magnetic field and claim the Google Lunar X Prize. Beresheet successfully completed its long travel—about two months—to the Moon but failed to land, and the prize is to be awarded to the first team to land on the Moon, drive 500 meters across its surface, and return pictures, and has not been claimed in 2018 for lack of contenders. There is still time for it. Morris Kahn, SpaceIL's president, declared that its team has already began the work on Beresheet

2.0; other companies are competing with the Israeli for the same prize, in what will be an interesting race to watch.

In 2019, some of the missions already on their way will enter in their crucial part of their work. NASA's InSight lander—which touched down in December 2018—has already begun the drilling into the surface of Mars and collect still-unknown data about the way rocky planets—like Mars and the Earth—have formed. In another daring mission, NASA's Parker Solar Probe will make its next close approach to the Sun in summer.

NASA is not going to steal all the spotlight: ESA's plans for 2019 are a series of fast turn-around, highly focused missions that complement the agency's medium-term Science Programme. One of them will be the launch of the planet-hunting mission CHEOPS (Characterising Exoplanets Satellite), which will look for planets orbiting bright stars close to our Solar System. It will complement the already working NASA-TESS on chasing exoplanets.

Looking further, the 2020s promise to bring astonishing accomplishments, with humankind coming back to the Moon, more space stations orbiting the Blue Planet, and maybe, a crewed mission to Mars. Many, if not all, of these missions will see partnerships between private and public sector along the lines that have been emerging in the last decade and that look the best solution possible to take the space industry as a whole to the next level. This will be particularly important in those segments—such as space tourism or space mining—where the private initiative has taken the lead since the onset.

Another important development, which looks essential to achieve an ambitious-yet-technical-feasible Moon Station many space agencies have declared as their long-term plans, will be an international level of cooperation to make them possible. The ISS would have not been a reality without it, and neither will be what follows. And if China has been excluded from the concept of International Space Station back in the 1990s, nobody would seriously think about not involving China in a lunar station nowadays, especially considering the kind of experience China is stocking up with its Chang'e missions and their exploration rovers.

The ongoing talks between NASA and the CNSA regarding cooperation on future Moon missions is a good way to start. But any more ambition plan will require a status of world politics that looks more positive than it does today. The creation of a US Space Force, which has been received with concerns by the rest of the world, has sent mixed signals regarding the willingness of the dominant space power to engage in joint initiatives.

Moreover, the ongoing trade disputes between China and the USA do not help to create a climate of confidence, which is essential for cooperation in a sector where security issues and military implications are always present. Nowhere as in space, the needle between conflict and cooperation needs to be as close as possible to the peaceful side for great achievements to come.

Timeline of Space Exploration

October 4, 1957:	First launch of an artificial satellite, Sputnik I (Soviet Union)
October 4, 1957:	First animal in orbit on Sputnik 2 (Laika, Soviet Union)
January 2, 1959:	†First artificial object to leave Earth orbit (Luna 1, Soviet Union)
*August 7, 1959:	First photograph of Earth from orbit (Explorer 6, USA)
April 12, 1961:	First man in orbit on Vostol 1 (Yuri Gagarin, Soviet Union)
May 5, 1961:	First suborbital crewed flight (Alan Shepard Jr., USA)
July 10, 1962:	First commercial satellite (Telstar 1, on a US rocket)
August 27, 1962:	†First successful interplanetary flyby (Mariner 2, Venus, USA)
September 29, 1962:	†The first satellite launched from a country other than the United States or Soviet Union (Canada's Alouette 1, from a NASA Thor-Agena B rocket)

S. Paladini, *The New Frontiers of Space*,
https://doi.org/10.1007/978-3-030-19941-8

*June 16–19, 1963:	Valentina Tereshkova, the first woman in space (Soviet Union)
March 18, 1965:	First EVA (Aleksei Leonov, Soviet Union)
November 26, 1965:	First successful French satellite (Astérix) launch
March 1, 1966:	First landing on another planet, but no data returned (Venus; Venera 3, Soviet Union)
April 3, 1966:	†First probe to enter lunar orbit (Luna 10, Soviet Union)
*June 2, 1966:	†First successful U.S. soft landing on the Moon (Surveyor 1)
January 27, 1967:	First casualties in a space mission (Gus Grissom, Edward White and Roger Chaffee, Apollo I, USA)
*December 21, 1968:	First manned spacecraft to orbit moon launched by a Saturn V (Apollo 8, USA)
*July 20, 1969:	Man walks on the moon (Neil Armstrong and Edwin "Buzz" Aldrin, Apollo 11, USA)
February 11, 1970:	First successful Japanese satellite (Ohsumi) launch
April, 24 1970:	First successful Chinese satellite (Dongfanghong-1) launch†on China's† first† Long March rocket (LM-1)
November 17, 1970:	First Lunar Rover
*April 19, 1971:	First space station (Salyut 1, Soviet Union)
*October 28, 1971:	First successful British satellite (Prospero) indigenous launch
November 14, 1971:	First orbit around another planet (Mars, Mariner 9, USA)
*December 2, 1971:	First soft landing and mission on Mars (Mars 3, Soviet Union)
*March 3, 1972:	First human made object sent on escape trajectory away from the Sun, passing the Asteroid Belt and leaving the Inner Solar System (Pioneer 10, USA)

May 14, 1973:	Skylab I, first U.S. orbiting laboratory, launched
December 3, 1973:	First Jupiter flyby (Pioneer 10, USA)
*March 29, 1974:	†First spacecraft to fly by Mercury (Mariner 10, USA)
July 17–19, 1975:	U.S. astronauts and Soviet cosmonauts participate in Apollo-Soyuz Test Project, docking together in space for two days
April 19, 1975:	†First successful India's satellite (Aryabhatta) launched by a Soviet rocket
*May 31, 1975:	†Creation of the European Space Agency (ESA)
July 20, 1976:	†First successful Mars lander mission (Viking 1, USA)
*August 20, 1977:	First spacecraft launched towards Uranus and Neptune (Voyager 2, USA)
September 5, 1977:	The twin spacecraft of Voyager 2 launched to perform flybys of Jupiter and Saturn (Voyager 1, USA)
*February 22, 1978:	†The first GPS satellite, Navstar 1, launches aboard an Atlas F rocket
February 22, 1978:	†The first GPS satellite launched (Navstar 1, USA)
September 1, 1979:	First spacecraft to fly by Saturn (Pioneer 11, USA)
December 24, 1979:	†Europe's first rocket vehicle launch (The French-built Ariane)
July 18, 1980:	India's first rocket vehicle launch (SLV-3)
April 12, 1981:	Space Shuttle Columbia becomes first winged spaceship to orbit Earth and return to airport landing
*June 18, 1983:	Sally Ride becomes first American woman in space on SST Challenger
February 7, 1984:	First untethered spacewalk off the Challenger space shuttle with a Manned Maneuvering Unit (USA)

September 11, 1985:	First comet flyby (International Cometary Explorer, USA)
*January 24, 1986:	†First (and only to date) spacecraft flyby of Uranus (Voyager 2, USA).
January 28, 1986:	Space Shuttle Challenger explosion 73 seconds after launch, killing its crew of seven
*February 20, 1986:	†First modular space station launch (MIR, Soviet Union)
September 19, 1988:	†First Israeli satellite (Ofeq 1) on an Israeli Shavit rocket
November 15, 1988:	Unmanned Space Shuttle Buran first (and only) launch (Soviet Union)
May 4, 1989:	†Magellan spacecraft launch from Shuttle to radar map the surface of Venus (SA)
August 25, 1989:	First (and only to date) spacecraft flyby of Neptune (Voyager 2, USA)
October 18, 1989:	†Jupiter-bound Galileo spacecraft launch from Shuttle
February 14, 1990:	First photograph of the Solar System (Voyager 2, USA)
April 25, 1990:	†Hubble Space telescope launch
April 23, 1992:	First detection of Big Bang's residual radiation (Cosmic Background Explorer, USA)
January 26, 1995:	†China's Long March rocket failure (that destroyed an Intelsat satellite) resulted in the U.S. government ban of American-built satellites on other countries' carriers (including China)
*March 14, 1995:	Norman Thagard becomes first American to be launched on a Russian rocket and to visit MIR
*June 29, 1995:	Atlantis Space Shuttle docks with Mir in first shuttle-station hookup
*July 4, 1997:	Mars Pathfinder (and its rover Sojourner) touch down on Mars (USA)

*November 20, 1998:	First multinational space station (ISS) launch. It is largest man-made object built in space to date (Russia, USA, ESA, JAXA, CSA)
May 29, 1999:	First space shuttle docking to the ISS (Discovery, USA)
*July 23, 1999:	†Shuttle launches Chandra X-ray observatory (USA)
*February 14, 2000:	First orbiting of an asteroid (Eros, NEAR Shoemaker, USA)
November 2, 2000:	First crew on the ISS (USA and Russia)
February 12, 2000:	First landing of an asteroid (Eros, NEAR Shoemaker, USA)
May 6, 2001:	†Dennis Tito (USA) aboard a Russian Soyuz spacecraft visit the ISS and become the world's first space tourist
February 1, 2003:	Second Space Shuttle accident (Columbia, USA)
†*October 15, 2003:	First Chinese manned spaceflight, and third country in human spaceflight to date (Yang Liwei on Shenzhou 5)
January 4 and 25, 2004:	†The first twin Mars Exploration Rovers, Spirit, and Opportunity lands on the Martian surface (USA)
July 1, 2004:	First orbit of Saturn (Cassini-Huygens, USA)
*January 14, 2005:	†Huygens probe lands on Saturn's moon Titan, becoming the first spacecraft to explore another planet's moon (USA)
*September 27, 2007:	†Dawn, the first ion-powered probe to visit two celestial bodies, launched towards steroid Vesta and dwarf planet Ceres in the Asteroid Belt
October 7, 2007:	First Chinese lunar orbiter (Chang'e 1)
*November 20, 2008:	First Indian lunar orbiter (Chandrayaan-1)
March 6, 2009:	Kepler exoplanet hunter telescope launch (USA)

†June 13, 2010:	First sample return from asteroid†(Hayabusa 1, Japan)
*March 18, 2011:	First orbit of Mercury (Messenger, USA)
July 21, 2011:	Final space shuttle mission landing (Atlantis, USA)
August 25, 2012:	First manmade probe in interstellar space (Voyager 1, USA)
*December 14, 2013:	China's first moon landing (Chang'e)
*September 24, 2014:	First Indian interplanetary orbiter; MOM is the first nation's probe to succeed at its first attempt to Mars
November 12, 2014:	First man-made probe to make a planned and soft landing on a comet†(ESA-Rosetta and its lander Philae)
*March 6, 2015:	First orbit of dwarf planet (Ceres, Dawn, USA)
July 14, 2015:	First Pluto's flyby (New Horizons, USA)
*September 15, 2017:	Cassini spacecraft plunged into Saturn's atmosphere in the Gran Finale, ending two decades of space exploration
January 1, 2019:	First flyby of a Kuiper Belt's object (Ultima Thule, New Horizons, USA)
*January 3, 2019:	First soft landing on the far side of the Moon (Chang'e 4, China)

Index

S. Paladini, *The New Frontiers of Space*,
https://doi.org/10.1007/978-3-030-19941-8

FSC
www.fsc.org
MIX
Papier | Fördert gute Waldnutzung
FSC® C083411

Kolibri 360 GmbH
Ferdinand-Jühlke-Straße 7
99095 Erfurt, Deutschland
produktsicherheit@kolibri360.de